PLANET IN DISTRESS

Environmental Deterioration and the Great Controversy

To order additional copies of *Planet in Distress,* by Scott Christiansen,
call 1-800-765-6955.
Visit us at www.reviewandherald.com for information on other
Review and Herald® products.

PLANET IN DISTRESS

Environmental Deterioration and the Great Controversy

SCOTT CHRISTIANSEN

REVIEW AND HERALD® PUBLISHING ASSOCIATION
Since 1861 | www.reviewandherald.com

Review and Herald® titles may be purchased in bulk for educational, business, fund-raising, or sales promotional use. For information, e-mail SpecialMarkets@reviewandherald.com.

The Review and Herald® Publishing Association publishes biblically based materials for spiritual, physical, and mental growth and Christian discipleship.

The author assumes full responsibility for the accuracy of all facts and quotations as cited in this book.

Texts credited to NKJV are from the New King James Version. Copyright © 1979, 1980, 1983 by Thomas Nelson, Inc. Used by permission. All rights reserved.

This book was
Edited by Lincoln Steed
Copyedited by Jeremy J. Johnson
Cover designed by Mark Bond
Cover photo © Thinkstock.com
Typeset: Bembo 11/13

PRINTED IN U.S.A.

16 15 14 13 12   5 4 3 2 1

Library of Congress Cataloging-in-Publication Data
    Christiansen, Scott.
      Planet in distress : environmental degradation and the great controversy / Scott Christiansen.
        p. cm.
     1. Human ecology—Religious aspects—Christianity. 2. End of the world. 3. Seventh-day Adventists—Doctrines. I. Title.
    BT695.5.C495 2012
    261.8'8—dc23
                  2012006049

ISBN 978-0-8280-2660-4

For my good friend,

Allen Darnell, who got me thinking about how both the character of God and the character of Satan are reflected in the systems and functions of nature as we know them; and how, before the taint of sin, all of creation may have functioned in a cooperative, synergistic fashion. It was from those thoughts that this book grew.

Contents

Prologue

Being raised an Adventist, I was exposed from an early age to the Bible, including the prophecies in Revelation and how they were and are coming to fruition. Even as a boy I could, at a very basic level, understand some of the fundamentals underlying what would happen during the last days. I more or less grasped, for instance, that sin would mean that humanity would become more and more evil and that human beings would kill other human beings. This was ingrained in me, for one of my earliest memories is of my mother standing at the ironing board, screaming, and dropping the hot iron on the floor when the news came over the radio that President Kennedy had just been shot. I still remember, with perfect clarity, the tendrils of smoke rising around the iron and the smell of burning nylon as my mother listened, horrified and stricken, to the news report. As I grew I added to this the daily staple of television reports on the Vietnam War, complete with their grim death statistics. I still remember my parents letting out gasps of pain and despair as the monthly death totals from that terrible war were announced. Even as a sheltered boy in an Adventist home I well knew that the world was a sinful place and that humans could—and did—kill humans.

As I grew, I also came to understand that the institutions of humanity could be and were used for evil, and that sin and corruption pervaded these institutions from the bottom to the top. I was only 12 years old at the time, but I clearly remember lying on my grandmother's living room carpet and watching the Watergate hearings on TV. Even at that young age, or perhaps especially because of that young age, I understood that humanity was essentially corrupt, selfish, and willfully sinful. The point was particularly clear to me as I had, only a couple years earlier, managed to weave my way to the front of a large crowd at the dedication of the new Loma Linda University Medical Center and had twice shaken the hand of President Richard Nixon, who had arrived in a helicopter to give the dedication speech. I still remember how, when I shook his hand for the first time, he had been sweating in the California sun, which made him seem very human. I then threaded my way through the crowd and inserted myself a little farther along the rope line to again shake his hand, and when he got to me his hand was bleeding from the nails of some over-eager citizen. Seeing a president sweat and bleed had made him seem human; seeing him later accused of high crimes and driven from office made all human beings seem more corrupt.

And with this view of humanity, the prophecies in the Bible and the comments of Ellen White made sense to me; I understood that the consequence of sin in humans was that humans would continue to grasp selfishly for power and gain and would cause pain and death until most of humanity, which had been created in the image of God, had been corrupted into the image of Satan.[1]

And if the selfish and hostile and corrupt relationships that humans had with humans were understandable to me, so too were the hostile relationships that human beings had with animals, and the fear and dread[2] that animals had for humans. I had, of course, seen a great deal of death in animals even before my teens. At that time, living in what was still a relatively rural Loma Linda in southern California, my father regularly dispatched gophers and rattlesnakes around the house. There were also the baby birds that had fallen from trees that my brothers and I tried to rescue and raise, with very few successes. And there were pets that died from natural causes and a good number of dead animals at the roadside—"road kill," as we so casually called it. The examples of death in animals were all around me, and I could understand why animals would be afraid of humans, and how sin in humans and the instructions of God could cause that fear (see Gen. 9:2).

But there was one thing that I heard said by parents and pastors and teachers that I could not understand. They told me that the whole world "groaned under the weight of sin." How could that be? As one writer clearly put it in commenting on Romans 8:22: "It is because of man's sin that 'the whole creation groaneth and travaileth in pain together.'"[3] This I just didn't understand. How could sin affect the actual, physical world? Could sin directly affect the soil and the trees and the mountains and rivers and oceans? I understood that humans would destroy humans and that humans would destroy animals—after all, I had seen that or at least been told about it, so I could grasp it. But I could not understand how sin "burdened" the entire world, and I didn't hear any more than very general explanations in response to the question.

At that time (going back to the 1960s and early 1970s) Loma Linda and the surrounding area was made up mostly of orange groves—fragrant and productive orange trees carpeted the land. Try as I might, I didn't understand how sin in humans could make the orange groves go away. Or make the snow and ice on mighty mountains melt. Or cause a storm. To me, there was a distinct difference between humans killing humans in anger or greed, or humans killing animals with glee or indifference, and the sin in humans killing some, or all, of the natural world. The only way I could see for that to happen was through a nuclear bomb (and the fear of such a bomb was high at that time in history), but even then I knew that were a nuclear bomb or bombs set off, much of the natural world would escape untouched.

As I surveyed the mighty snowcapped mountains that surround Loma Linda, as I walked in the fields and orange groves and dry riverbeds close to my house, and as I observed the overwhelming abundance of life in every nook and cranny around me (everywhere were spiders and bees and wasps and worms and moths and bugs and flies and mites—a paradise for a curious boy!), it just seemed to me that, while human beings could kill animal life, the animal life for the most part went on without taking much notice of him. I could see no relationship between sin in humans and the destruction of the earth, nor could I see that it was even possible for sin in humanity to destroy the world. After all, it says in Genesis 1:31: "And God saw every thing that He had made, and, behold, it was very good." The world, which God Himself thought was "very good," was, in my mind at least, simply too big and complex and well balanced for sin to impact it. Sin, to my thinking, was something found in humans, not something found in nature.

As I grew older I had more experience with the good and evil that is in humanity, and my opinion of humans evolved correspondingly. As I matured I also traveled to different places and different climates—as a missionary to the Navajo Indians at La Vida Mission in the high mesas of New Mexico, and as a witness to the people on the frigid steppes of Mongolia while working as country director for the Adventist Development and Relief Agency (ADRA). After Mongolia I was able to work in the dynamic but toxically polluted China, again as country director with ADRA.

But while I was in Mongolia I began to learn (and most of the world began to learn with me) about global warming and climate change. While there was a great deal of debate and skepticism about climate change at that time, I was able to see its effects firsthand once they were pointed out to me. I saw the deserts advancing year to year in Mongolia and causing displacement of families,[4] and I saw the devastating winter snows *(zuud)* that were without precedent in Mongolia and that caused hundreds of thousands of nomadic herders to lose all their livestock and end up in hopeless poverty. The Mongolian herders didn't cause global warming (indeed, they use almost zero nonrenewable carbon), but they were profoundly impacted by the emissions of a world selfishly focused on making more money and acquiring more luxury. It was in Mongolia that I began to get an inkling of how the cumulative effect of sin on the natural and physical systems of the world could actually cause degradation and the eventual collapse of something as vast and complex as the global systems that God created in perfect and elegant balance.

The evolution of my thinking on the degradation and collapse of global systems continued when my ADRA duties moved me from Mongolia, one of the least-densely populated countries on the planet, to China, with its teeming 1.3 billion people. With the mind-boggling number of people in China, each engaged in competing for and exploiting a niche, and the sheer density and extent of the industrialization in the country, the degradation of the natural systems that the Chinese rely upon for life was shockingly evident to me. Between China and Mongolia, I was able to witness firsthand the significant effect that sinful individuals and towns and states and nations are having on the natural systems that God put in place with perfect balance, and how those systems are in decay on a global basis.

But even though I had seen the degradation of some of the world's systems firsthand, it was not until I returned to the United States, settled in Maine, and began a purposed reading of significant texts, inspired writings and scientific reports (many of which are referenced in this book) that I began to see the interconnection between global systems (my own clumsy phrase for the worldwide systems that God created that underpin life on earth, some of which are covered in this book) and the selfish and sinful actions of humanity. And it was not until I was in my late 40s and working on my master's degree that I began to really see both the interdependent nature of global systems (both those created by God and those that are human-made) and the increasingly apparent decay of those systems.

But it was only when I explored the link between Satan's usurpation of the earth and his inability to control and maintain the perfect balance and order that God created in the systems of the earth that I realized that the primary effect of sin—decay and deformity and death—is a direct result of God's creation being separated from God. And it is in this light that I now see the decay of the systems of the earth as a direct result of the combined effect of their separation from God and the effects of Satan's malignant administration of the earth.

The implications of the slow decay in global systems, which is now becoming evident, are staggering. The implications attending the current acceleration of decay are mind-boggling. Simply put, as even one global system rapidly decays, and especially as multiple global systems decay, billions of people will be thrown into dire poverty and much worse, and the societies of the world will not be able to function as they have been—with the result that the nations of the earth will be angry.

While I studied and absorbed the wisdom of humanity on the decay of various global systems, I almost automatically worked in parallel to relate what I was learning with what I knew from the Bible and from the inspired writings of Ellen White. When I returned to school for my master's degree, I at the same time accepted a position as lay pastor to the charming and very small Adventist church in Harrison, Maine, and on more than a few Sabbaths those patient souls had to sit through longer-than-usual and more-technical-than-usual sermons either directly on the subject of global system decay (and what it means to Adventists) or indirectly on the same subject. As I studied and preached, I began to grow in my conviction that the destruction by sin of the global systems that we depend on for life on this planet is related to the birth pangs that signal the coming of the time of trouble. This realization impacted me and depressed me profoundly, but only briefly. For after a while I realized that it is not *what* is coming, but *who* is coming that should be the focus of my thoughts and actions. Using this perspective as a springboard, I grew to see that the social upheaval attending global system decay will provide faithful Christians everywhere with an amazing opportunity to minister to others and to spread the word of Christ's soon coming with great power and effectiveness—but only if we can struggle out of our Laodicean mind-sets and only if we can withdraw our emotional attachment to the social and economic systems that are increasingly in danger of collapse because of the stresses and pressures of global system decay.

With that context I offer this book as my thoughts on what is coming, and my thoughts on how to use the approaching events and situations to work powerfully for God and prepare the teeming throngs on earth for Christ's coming. Researching this book has been a powerful spiritual experience for me, and it is my hope that you will read this book critically and prayerfully, and, if God leads you to come to the conclusion that some parts of this book are useful and instructive to you, it is my prayer that the Holy Spirit will urge you to solemnly assess the situation in the world, then plan and take truly vigorous action for God in these waning days. With this in mind, I have split the book more or less into three sections. The first section of the book frames the discussion using the Bible and inspired writings. The second section deals with the description of global systems and the forces that science and worldly learning see at work in their decay. This section of the book has been somewhat more liberally referenced and endnoted, and I urge the skeptical or curious to go to the original sources and also to seek out significantly greater coverage of specialty topics. This is offered with the caveat that our knowledge of the earth and its systems (and their decay) is rapidly progressing because more scientists are now studying the earth than have ever studied anything before, and thus knowledge that is just a few months old may be outdated. Therefore, those interested in earth and climate science will want to study well beyond the sources presented here.

The third and final section of this book deals with what the coming events mean to Christians and especially to Adventists, including how we can begin to pull away from society (which depends wholly on the systems that are in decay) while maintaining our witness to society, and a few things we can do to buffer the impact that is coming. But more than anything else, the third section of the book concentrates on the parallels in Scripture between the effects of global system decay and the state of the world as it will be just before the time of trouble. In this sobering context, the final section of the book underscores that we should be working to prepare our hearts and the hearts of others for the soon coming of Christ, and that we should prepare to work in power at the pouring out of the Holy Spirit—a pouring out that I believe has already begun.

A caveat: When my parents got married, they initially discussed the possibility of not having any children because of their sincere belief that the coming of Christ was so close. As sometimes happens, they instead had four boys in relatively short order. When my wife and I were engaged, we also discussed not having children because Christ's coming was so close. As it happened, the first of our four children was born 11 months after we were married. As I write this book, that baby is now a strapping young man who is a first-year medical student at Loma Linda University. And so, if we apply human time scales, my parents and I were both wrong. The result of being wrong, though, is that both my parents and my own family lived lives that were more urgently dedicated to God. Now, it might be that the research and thinking I have done in preparing this book is in error. Or it might be that the work of other humans upon whom I have relied is in error or simply premature. However, if that is the case, I see no harm. This book urges Christians to look at the palpable nearness of the coming of Christ, to make their lives and their hearts right with God, to dedicate themselves to His service, and to evangelize others with urgency and passion. Such actions, if taken, will save many souls and draw many Christians yet closer to God. If I am too early with my message by 10 years or 100 years, the result still glorifies God and swells His kingdom. So, while my work has been both careful and prayerful, if I am wrong and it causes even one soul to be won for Christ, then it is worth both the labor and embarrassment of being too early.

Finally, a word about fanaticism: This book is not a manual for developing strategies to survive the coming crises. This book will not help anyone figure when it will be best to sell their home, or convert their assets into gold, buy survival gear, or any other such foolishness. Everything belongs to God and, if He wants any of it saved or moved around or sheltered, He will let His faithful children know, if they will but seek His voice. People who are interested only in survival strategies or fleeing immediately to the hills will not find this book rewarding and should turn elsewhere for their reading. People who love Jesus, who are concerned about the salvation of others, and who sincerely want to prepare the way for the Lord, may find this book useful. Such is my prayer.

---

[1] We are told in *Patriarchs and Prophets* (pp. 78, 79) that humanity, just prior to the Flood, was a reflection of Satan's administration of the earth. If this happened in a relative few generations, how more so are human beings today remade in Satan's image?

[2] Ellen G. White, *Patriarchs and Prophets* (Mountain View, Calif.: Pacific Press Pub. Assn., 1890), p. 107.

[3] *Ibid.*, p. 443.

[4] Once the sparse grass was killed by the advancing desert, families who relied on their animals for food, fiber, and fuel were forced to move.

Chapter 1:

In the Beginning

"And God saw every thing that He had made, and, behold, it was very good." Genesis 1:31.

I did not begin to appreciate just how invested a creator is in the thing that is created until I started writing as a young man and found that I was very careful with (and possessive of) the things I created. Later, when my wife and I had our first child, I was amazed at the little life that we had created, and fascinated by everything our baby boy was capable of doing. How amazing it is that our Creator has given us the ability to create—and with it the joy of being a creator—and how like our Creator to want to share this privilege. However, even though we can create, we probably will be able to grasp only a little of what was going on when God created heaven and earth. This is because God created everything out of nothing, and did so using just His spoken word. We cannot readily conceive what it is like to create something out of nothing—we simply have no frame of reference. Still, because we were created in the image of God and because He gave us some measure of His characteristics, we can through study grasp a little of what went on that week and feel awe for the world as it was, the world that all of heaven sang and rejoiced over when it was finished.

In the beginning God created the heaven and the earth. And it was an amazing, astonishingly beautiful, and perfectly balanced paradise. The Bible gives only a scant account of the Creation week—how with the spoken word God created the earth, the sun, the moon and the stars—and we thus have only a mere summary of the work God undertook in the six days before He rested on the seventh. But just the summary we are given is amazing, and as man grows to know more and more about the intricate and finely balanced systems that God put in place during Creation, amazement grows to astonishment. Or at least it does for Christians who accept a literal Creation week. Humanity's understanding of God's creation, while it has grown tremendously, is still very superficial. But from what we do understand, it is very apparent that God created interlinked and interdependent systems of enormous complexity that functioned as a perfect whole to maintain the earth as God created it. Few Christians, it seems, stop to ponder deeply what happened during Creation week, or what has happened to God's creation since that week, or how the earth itself—the actual planet and its life and energy systems—figures into the war between Christ and Satan. This is a pity, since meditating on these things can deliver deep spiritual blessings.

Open your Bible to Genesis 1 and give a considered reading to that chapter. The chapter starts, fittingly, with a beautiful and poetic description of what things were like before creation: "And the earth was without form, and void; and darkness was upon the face of the deep. And the Spirit of God moved upon the face of the waters" (Gen. 1:2). Next, God spoke. With His spoken word He called into existence light: "And God said, Let there be light: and there was light. And God saw the light, that it was good: and God divided the light from the darkness. And God called the light Day, and the darkness he called Night. And the evening and the morning were the first day" (verses 3-5). Here we see that God created light *before* He created the sun, the moon, and the stars. God also created the first day, by separating the light from the darkness. Thus, on the first day, God created some crucial systems—He created both light and time and intertwined them (Einstein has a great deal to say about the intertwined nature of light and time), not least by marking the passage of the day with light. God also created the first day of a calendar that has continued to our time, and He created both energy and the transport system for that energy (light). In creating light, time, and our calendar, God created critical systems on the first day of the Creation week—systems that continue to affect and regulate the lives of every human every day. And by creating light and time, God seems to have created most if not all of the laws of physics by which the universe operates—in other words, at the start of Creation it seems He created the natural laws that govern His creation. This expanded view of the first day of Creation, with God laying the foundation for the rest of His work and putting eternal rules in place, seems more fitting when compared to God "just" creating light on the first day of Creation.

But there is another thing that happened in these few verses that does not get much attention: note that God assessed His work. In the first part of verse 4 we read: "And God saw the light, that it was good." This verse begs a question: How good is "good" for God? Given what we know of God, it is reasonable to draw the conclusion that what He created was perfect, in perfect conformity with His intentions, and was meant to last forever. What is "good" for God is probably beyond the comprehension of humans living in a sinful state, and it seems we can safely substitute our word "perfect" for God's "good."

On the second day of the Creation week God created a massive and amazing system—the first of a matching pair—that all humans and much of the rest of the forms of life on earth rely upon: our atmosphere. In Genesis 1:6-8 we read: "And God said, Let there be a firmament in the midst of the waters, and let it divide the waters from the waters. And God made the firmament, and divided the waters which were under the firmament from the waters which were above the firmament: and it was so. And God called the firmament Heaven. And the evening and the morning were the second day."

Speaking our atmosphere into existence was no small thing. The gases that surround the earth and interact and buffer each other are enormously complex and are made of discrete layers, with each successive layer being thinner and thinner, until our atmosphere gives way to space itself. Without these layers of gases, arranged in their exact order and thickness and each with its varying function, everything on earth would rapidly die. This is because, aside from the fact that most life on earth needs oxygen to breathe, the atmosphere also protects us from harmful radiation and traps just the right amount of the warmth of the sun, keeping the temperature on earth within a livable range. Alas, the atmosphere as God created it had a different design than the one we have—in *Patriarchs and Prophets* we read that it was designed to maintain a steady temperature[1] and to function without raining. And of course, it was designed to keep working in perpetuity. To achieve these functions, the mixture of gases that surround the earth had to be precisely correct, and had to be designed with buffering systems and other checks and balances. Perhaps it is because these other checks and balances were not yet in place that God did not assess His work at the end of the second day, but instead waited until the third day—when He was finished creating the seas, which are the mirror image of the atmosphere and which interact with the atmosphere and exchange gases, thus stabilizing both systems.

In Genesis 1:9, 10 we read: "And God said, Let the waters under the heaven be gathered together unto one place, and let the dry land appear: and it was so. And God called the dry land Earth; and the gathering together of the waters called he Seas: and God saw that it was good." Thus on the third day of Creation God created the seas. This was more than just putting a great deal of water in one place. The seas have in them an extraordinary collection of minerals, salts, and nutrients that help sustain the myriad of life-forms in them, and currents in the seas flow ceaselessly around the earth and through the depths, collecting and distributing elements and nutrients throughout the seas. God designed that the balances of these minerals and elements in the water be extraordinarily precise and would exactly suit the needs of the creatures He would create later in the week. And He would also create an array of subsystems that would buffer the seas, keeping nutrients and minerals (and gases and acids) within their intended bounds.

But the function of the oceans was not just about balancing and moving about minerals and salts and elements—the seas were also designed as a thermal regulator for the planet (a job shared by the atmosphere as God created it). This particular function of the ocean involves the heating of water in the tropical seas, and the flowing of that water toward polar regions. By this means heat is distributed around the world, and a greater degree of thermal stability is achieved. The description of the functions and complexity of the oceanic systems provided here just barely scratches the surface, but it begins to convey the magnitude and importance of this system that God created on the third day of Creation week. After God finished creating the seas, He looked at His work and pronounced it "good." In other words, everything about the seas and the ways in which they achieved His purposes was perfect, in perfect balance, and would stay that way for eternity under His care. And by extension, the matching half of the waters that were divided—the firmament that was created the day before—was also perfect.

But God was not yet ready to end His third day of Creation. Having created dry land when He gathered the waters together, He went on to create all the plants that are on the earth. In Genesis 1:11-13 we read: "And God said, Let the earth bring forth grass, the herb yielding seed, and the fruit tree yielding fruit after his kind, whose seed is in itself, upon the earth: and it was so. And the earth brought forth grass, and herb yielding seed after his kind, and the tree yielding fruit, whose seed was in itself, after his kind: and God saw that it was good. And the evening and the morning were the third day." And so it was that on this day—afternoon, really—God created every green thing on the face of the earth. His creation that afternoon covered an amazing range of plant life-forms, from the microscopic, such as algae, to the truly massive, such as the sequoia redwood. The green layer that suddenly covered the planet interacted with the atmosphere (and therefore, indirectly, with the seas), consuming massive amounts of carbon dioxide (part of the "carbon cycle") and emitting staggering amounts of oxygen (green plants taking in light, moisture, and minerals to manufacture sugars used by them to grow while discharging oxygen as a waste product is called "photosynthesis"). But an earth full of plants needed more than carbon dioxide to grow, and thus the nitrogen cycle was initiated (where nitrogen is precipitated or otherwise "fixed" out of the atmosphere and cycles through plant and water life) and minerals (particularly alkali metals such as potassium, magnesium, and calcium) began cycling through the plant systems and through water systems. Just the beginnings of what God created was amazing, and all of it meshed together and was in perfect balance.

On the fourth day of Creation week, recounted in Genesis 1:14-19, God made the sun, the moon, and the stars. "And God said, Let there be lights in the firmament of the heaven to divide the day from the night; and let them be for signs, and for seasons, and for days, and years: and let them be for lights in the firmament of the heaven to give light upon the earth: and it was so. And God made two great lights; the greater light to rule the day, and the lesser light to rule the night: he made the stars also. And God set them in the firmament of the heaven to give light upon the earth, and to rule over the day and over the night, and to divide the light from the darkness: and God saw that it was good. And the evening and the morning were the fourth day."

We know that from the first day of Creation, God created a calendar that is still in effect. But it was not until the fourth day that He created the solar system that regulates His calendar, while also regulating the seasons. Few systems are as fundamental to life on the earth as the sun and moon— the sun constantly creates massive amounts of energy, a small portion of which is received by the earth. A still smaller portion of the energy the earth receives from the sun is trapped in our atmosphere, creating a livable climate. Very small amounts of energy (relative to what is created by the sun) also drive our ocean currents and our winds in the upper and lower levels of the atmosphere, which, through complex interaction with the seas and land masses, is what drives our storms and rain patterns. In addition, the energy from the sun is used by plants as they use carbon dioxide and produce oxygen. And so it was on the fourth day that God created the energy that drives most of the key systems and processes on earth—plant growth, nutrient exchange, nitrogen and carbon cycles, gas exchange between plants and atmosphere, and between atmosphere and oceans, and climate stability. It is possible that there were still other functions that the moon had that are lost to us now. We know that before the Flood there was no rain, and that "there went up a mist from the earth, and watered the whole face of the ground" (Gen. 2:6). There may have been some interaction between the moon and the mist. It is interesting to note that the most ideal plant growth systems designed—hydroponics—operate on a cyclical watering system close to what the Bible describes existed on earth before the Flood. What we do know is that the gravitational pull of the moon creates the tides in the seas, a critical function in the life of the oceans. It may have been that the gravitational pull of the moon activated the mist that watered the earth. We know that the crust of the earth was dramatically different after the Flood since "the fountains of the great deep [were] broken up" (Gen. 7:11). But this is all speculation. What we *do* know is that God looked at what He had put in place and saw that it was good. In other words, it was perfect.

As the Creation week progresses, we note that everything that God created was dependent on or complemented what had been created the day before. God did not, for instance, create trees before He created dry land. Looking at the Creation week, we also see that on each successive day the systems that God created were more complex and more interactive, building on and relating to everything that was created before them.

On the fifth day God created all the fish and animals that live in the seas, as well as all the birds of the air. In Genesis 1:20-23 we read: "And God said, Let the waters bring forth abundantly the moving creature that hath life, and fowl that may fly above the earth in the open firmament of heaven. And God created great whales, and every living creature that moveth, which the waters brought forth abundantly, after their kind, and every winged fowl after his kind: and God saw that it was good. And God blessed them, saying, Be fruitful, and multiply, and fill the waters in the seas, and let fowl multiply in the earth. And evening and the morning were the fifth day."

It is interesting that, again, God treated the oceanic systems and the atmospheric systems—the "waters above" and the "waters below"—as two parts of a whole, in that He created the life-forms in the ocean and the birds on the same day. Once again we see that God's creation proceeded in an orderly fashion, in that the amazing array of sea life and birds that He created on the fifth day relied upon and interacted with the systems that

God had previously created. From the lowliest worm in the depths of the sea to the mighty blue whale (the largest creature ever known to exist[2]) and from the hummingbird to the ostrich, all of the creatures that God created on the fifth day needed the systems that He had previously created, forming chains of nutrients and energy that flow from the cycles powered by the sun and mediated and facilitated by a myriad of creatures and plants, all working in harmony to achieve God's design in His original creation.

Again we see that God assessed His day's work, and again He found it "good." How amazing that God created the unnumbered forms of life in the sea—as well as the birds—in one day and summed up His evaluation of His work with the simple word "good." If man makes a very good painting or creates a symphony or forms a union and creates life, they and the results of their work are showered with praise. But God did not allow Himself hyperbole, keeping His assessment to "good." And yet as good as it was—and it was perfect—God did not heap praise on His work. He was holding back, and He was doing it for a reason.

On the sixth day God had two major creation events. We read of the first of these events in Genesis 1:24, 25: "And God said, Let the earth bring forth the living creature after his kind, cattle, and creeping thing, and the beast of the earth after his kind: and it was so. And God made the beast of the earth after his kind, and cattle after their kind, and every thing that creepeth upon the earth after his kind: and God saw that it was good."

It is amazing that God created all land forms of life in a mere portion of a day. But for His final act of creation God would call into existence something very different—a form of life that looked like God, and was a free moral agent. All the other forms of life created by God were made without free will—without the choice as to whether or not they would trust and obey God and live within His laws. But on the final day of Creation, God created a life-form that would delight in His creation, would manage it, and, yes, could defile and destroy it with sin. We read in Genesis 1:26-28 that as His final act on the sixth day, God created human beings: "And God said, Let us make man in our image, after our likeness: and let them have dominion over the fish of the sea, and over the fowl of the air, and over the cattle, and over all the earth, and over every creeping thing that creepeth upon the earth. So God created man in His own image, in the image of God created he him; male and female created he them. And God blessed them, and said unto them, Be fruitful, and multiply, and replenish the earth, and subdue it: and have dominion over the fish of the sea, and over the fowl of the air, and over every living thing that moveth upon the earth."

And so God completed His great creation, but did so with one special instruction that gives us significant insight into just how different the sinless creation was from the world we live in today. We read that special instruction in Genesis 1:29, 30: "And God said, Behold, I have given you every herb bearing seed, which is upon the face of all the earth, and every tree, in the which is the fruit of a tree yielding seed; to you it shall be for meat. And to every beast of the earth, and to every fowl of the air, and to every thing that creepeth upon the earth, wherein there is life, I have given every green herb for meat: and it was so."

God made a special point of giving humans and animals plants to eat. In other words, everything on earth was created a vegetarian, and it would not be until long, long after humans sinned that God gave them specific permission to eat the flesh of animals (see Gen. 9:1-4). This fact gives us some small insight into how radically different the world was at Creation from the world we know today. We will later look at the changes in the world and God's systems since Creation, but first we will complete the first chapter of Genesis, for there is one verse left, Genesis 1:31, and it is a subtle yet critically important verse: "And God saw every thing that he had made, and, behold, it was very good. And the evening and the morning were the sixth day."

If what God created on the days of Creation were deemed by Him to be "good"—a description that we take to mean "perfect"—then how can a completed Creation be "very good," which would make it better than "perfect"? The clue lies in the fact that in the Creation week God assessed in turn each component of the earth that He made, each with its own systems, and found each perfect in turn and pronounced them "good." But it was not until the close of the Creation week that God assessed the earth as a whole ("And God saw everything He had made . . ."), which necessarily required that He assess the interaction of the systems He had made—how the atmosphere interacts with the ocean and with the plants and with the animals, and how the water is cycled through earth and vegetation and the atmosphere and seas, and how the energy flows from the sun to earth's various systems, etc. God looked at the perfect systems that He held in perfect balance—from single-celled life-forms in the seas and soil to the vast cycles that power the ocean currents and the jet stream to the vegetarian bears and lions—and declared the whole to be "very good." The earth as God created it must have been absolutely amazing—stunning—in its perfection and order and balance and beauty in order for God to declare it "very good."

Planet in Distress

The science behind the workings of the earth's systems is extremely complex and, frankly, very imperfectly understood by humanity. Since this book does not attempt to explain the science behind the earth's systems in detail, perhaps the best way to frame the discussion of the interconnectedness of earth's systems and the creation of those systems is to compare the Creation week to an extremely well-crafted gold pocket watch. Using this example, when God created the earth, He created the case of the watch. When He created the sun, He created the spring within the watch, since it is the sun that ultimately powers and drives almost all systems on earth. And when He created the various other systems that make the earth work, from plants and animals to seas and atmosphere and the biogeochemical[3] subsystems that support them, He was crafting the dozens and dozens of fine and delicate pieces of the workings of the watch. When God crafted each piece—whether case or spring or fine inner gear—He inspected it and called it good. But only when the watch was all put together, and the dozens and dozens of fine inner gears were meshing and interacting and driving each other in perfection, causing the watch to keep flawless time, did God call it "very good."

And it *was* very good—good beyond our ability to understand. And Adam and Eve were delighted, as God intended, by the earth and the Garden of Eden. But the state of perfection would not last long. Sin would enter the world and start a series of changes and decay and slow death to the systems of the world, and it is this decay and death of God's created systems that has continued to build through the ages and is culminating in our day.

---

[1] E. G. White, *Patriarchs and Prophets*, p. 61.

[2] See www.bbc.co.uk/nature/life/Blue_Whale.

[3] A big fancy word used to describe the chemical cycles on earth, and which recognizes that both the earth itself and the life-forms on the earth are an intricate part of these cycles. According to the *McGraw-Hill Science and Technology Dictionary*, the biogeochemical cycle is defined as "the chemical interactions that exist between the atmosphere, hydrosphere, lithosphere, and biosphere."

Chapter 2:

Sin and the Destruction of the World: Tracing the Arc From Beginning to End

"The sin of man has brought the sure result—decay, deformity and death. Today the whole world is tainted, corrupted, stricken with mortal disease. The earth groaneth under the continual transgression of the inhabitants thereof."—Ellen G. White, in The Seventh-day Adventist Bible Commentary, *Ellen G. White Comments, vol. 1, p. 1085.*

When sin entered God's perfect and perfectly balanced world, it changed everything. As selfish and self-centered humans, we tend to look at the effect of sin almost exclusively in how it impacts us, or at least in how it has affected humans in general, but in truth the impact of sin is much broader—there is no part of the world and no natural system in the world that has not been profoundly impacted by sin. This is a striking realization when paired with the thought that Satan wanted to have control of the universe,[1] and had God given Satan free rein, the vast universe would be in the same dramatically decayed, mismanaged state that the world is in today. The reason for this is simple—creation, when separated from the Creator, dies, and Satan cannot reverse or slow or otherwise mitigate this truth. In fact, it seems certain that Satan's selfishness-centered administration of the earth hastens the process.

When sin entered the world, the most immediate impact was the change of control. Where God had administration over the world that He created to His specifications, Satan eagerly took control. Ellen White writes that "not only man but the earth had by sin come under the power of the wicked one. . . . When man became Satan's captive, the dominion which he held, passed to his conquerer. Thus Satan became 'the god of this world.' 2 Corinthians 4:4."[2] The position of Satan as controller or "prince" of this world is reinforced in Scripture in Matthew 4:8, 9, where Satan offers Christ the world if only Christ will bow down to him. It is almost unnecessary to point out that Satan would not be able to offer the earth without first possessing it.

But of course, with sin and Satan's rule of the world came death, just as God had warned. Death did not come immediately to every living thing—various life-forms were impacted in varying ways and at varying speeds. We are told that parts of plants—such as flowers—were noted by Adam and Eve to be dying when they left the Garden of Eden,[3] and so we know that the impact of sin, through separation from God, was both immediate and significant. In *Patriarchs and Prophets* we read that "the harmony of creation depends upon the perfect conformity of all beings, of everything, animate and inanimate, to the law of the Creator."[4] With Satan usurping control over the world, and becoming "god" of the world, the world was in effect in rebellion and separated from God, thus assuring that even inanimate parts of creation were no longer in conformity with the law of the Creator.

In the previous chapter the completed creation was compared to a perfectly executed and intricate gold pocket watch, with fine inner workings. Returning to this metaphor: from the moment that sin was introduced, the watch began to corrode. The finest, most delicate workings were the first to go, and represent parts of creation that are lost to us as death, disease, and degradation has claimed untold plants, animals, and subsystems that we will not know of until we see and study the re-created earth in all its glory. It is true that the major pieces of the watch continue to grind away, but certainly not as God created or intended them, and even these major pieces (or major "systems," as this book refers to them) are winding down and beginning to fail from the accumulated decay that stems from sin.

Inspired writings affirm what we learn in the metaphor of the watch. In *Patriarchs and Prophets* we read: "Thus were revealed to Adam important events in the history of mankind, from the time when the divine sentence was pronounced in Eden, to the Flood, and onward to the first advent of the Son of God. . . . Crime would increase through successive generations, and the curse of sin would rest more and more heavily upon the human race, upon the beasts, *and upon the earth.*"[5]

The effects of sin were at work everywhere and on everything, and while flowers and leaves were quick to succumb, it would take hundreds of years before Adam and Eve would die. It would take thousands of years before the world itself would reach the decayed state described in Matthew 24. It would, in short, take until the present day before the massive and awesome systems designed by God to underpin the functioning of the world became so unbalanced that, like a still-spinning but slowing child's top, they would begin to wobble and tip out of control. We will look at the present-day effects of this more closely in chapters 4 through 7.

There was an immediate effect on the earth from the combination of the separation from God and Satan's poisonous control. There has been much discussion, and no small amount of argument, over climate change in our current age, and we will indeed look at climate change in greater detail in a later chapter. But it is worth noting here that climate change on earth under Satan's control was almost instant. In fact, the first noticeable impact of sin on the earth was climate change, and it was noticed on the very day Adam and Eve sinned. That bears repeating: the first noticeable result of sin was an almost immediate change in climate. We are told of Adam and Eve that, on the day of their transgression, "the air, which had hitherto been of a mild and uniform temperature, seemed to chill the guilty pair."[6] Now, perhaps there will be some who say that the "chill in the air" was a result not of climate change but of the loss of the "robe of light"[7] that Adam and Eve had worn prior to their fall. Fair enough. Here then is another comment that is far less ambiguous: "The atmosphere, once so mild and uniform in temperature, was now subject to marked changes, and the Lord mercifully provided them with a garment of skins as a protection from the extremes of heat and cold."[8] It was, in clear words, the "atmosphere" that changed as a result of sin. That is not a small point, and we would do well to stop and consider it: The introduction of sin (and the separation from God—the two are essentially two sides of the same coin) to the earth rapidly resulted in changes to something as massive as our climate system. The power of sin to decay and deform is that strong and that fast. We can only conclude that what God had created in perfection and in perfect balance began to degrade from the moment Satan assumed control of the earth.

We see that Satan had substantial control over the whole of the earth and its systems—not complete control, as he was and is restrained to an unknown degree by God,[9] but substantial control. However, while Satan aspires to the throne of God, Satan is not God, and even if he desired, he could not keep in perfect order and balance the amazingly complicated world that God created. A good way to think of this is by using a fish tank as an example. Anyone who has kept fish knows that even the simplest of setups—a goldfish in a bowl—is a complex system that can easily be upset. While an average person putting a goldfish in a bowl of water may not think about it at the time, what they are actually trying to do is set up a sustained system that balances the carbon cycle, the nitrogen cycle, gas exchanges between water and atmosphere, the mineral cycles, and the nutrient balance within the water, just to name a few. Fortunately for those of us who keep goldfish, we don't actually have to establish all these systems—we just have to set up our goldfish in such a way that microorganisms can get established in the bowl (on the rocks and the surface of the glass) that will digest and recycle the waste and excess food from the goldfish and thus keep the bowl in balance, facilitating most of the above systems. If we fail in this, by feeding too much too fast, for instance, or by putting too many fish in the bowl, or by putting the goldfish in a bowl without rocks or other surfaces for the necessary microorganisms to grow on, the result will be cloudy water followed not long after by the death of the entire system.

If earth is the largest and most complicated fishbowl ever devised, then Satan as usurper of that fishbowl is not qualified to keep it in perfect balance. That said, something as vast and complicated as the systems of the earth take time—in our case, thousands of years—to unbalance and reach the point of rapid degradation. But as with a fishbowl, the final stages of degradation, imbalance, and death come quickly, as we are beginning to see in our world.

Thus it is that the expected state of the natural world just before Christ returns (as described in Scripture and the Spirit of Prophecy) and the spectacular disruptions that are now being observed and that will grow greater and greater in our age are a direct and natural result of sin in the world, and a direct result of Satan's control of the earth. In other words, the line that can be traced from the first sin to the state of the natural world today is a clear and direct one. Note that the decayed and troubled state of the earth just before the rapidly approaching time of trouble and close of probation is not because of a random event or some whim of God, but is the natural result of sin—it is what happens when God's creation is separated from God.

Turn to Matthew 24:6-8 and consider what Christ predicted the state of the natural world would be just before His return: "And ye shall hear of wars and rumours of wars: see that ye be not troubled: for all of these things must come to pass, but the end is not yet. For nation shall rise against nation, and kingdom against kingdom: and there shall be famines, and pestilences, and earthquakes, in divers places. All these are the beginning of sorrows." Turn also to Isaiah 24:4, 5: "The earth mourneth and fadeth away, the world languisheth and fadeth away, the haughty people of the earth do languish. The earth also is defiled under the inhabitants thereof; because they have transgressed the laws, changed the ordinance, broken the everlasting covenant."

The understanding that everything negative and harmful that is happening to the earth and to the billions of people on the earth today is a direct and logical result of sin, from the first sin to the accumulated sin of the ages—and not the result of a vengeful or capricious God—is a key one, and is especially important for people who do not understand the nature of God and ask how He could cause disasters and suffering. God loves His creation (see the first six words of John 3:16), and God hates sin because it destroys His creation, to the point that Satan's rebellion, if left unchecked, would have destroyed the entire universe. The charge that God is the one who randomly destroys the earth and the people on it is one of Satan's most vile and effective deceptions, as it makes grief-stricken survivors of disasters blame God (who loves them and suffers with them) instead of blaming Satan, who caused their suffering and exalts in it. This key understanding is one of the messages that Christians need to convey to the world in these last days.

When God created the earth, it was a far cry from what we see today. As was noted in the previous chapter, rain did not exist, but instead the earth was watered with a mist that came up at night (Gen. 2:6). The trees were much larger, everything was far more productive, and nature operated more harmoniously. In a description of the world before the Flood, we read this: "The hills were crowned with majestic trees supporting the fruit-laden branches of the vine. The vast, gardenlike plains were clothed with verdure, and sweet with the fragrance of a thousand flowers. The fruits of the earth were in great variety, and almost without limit. The trees far surpassed in size, beauty, and perfect proportion any now to be found; their wood was of fine grain and hard substance, closely resembling stone, and hardly less enduring."[10]

Productive and beautiful as the earth may have been, not long after the fall of humanity the earth was resting under the first two of three eventual curses from God. Each of these curses suppressed the productivity of the earth and fundamentally changed the relationship between humanity and the earth. It is important to note that for this to happen seems to require that the very systems that God created in Creation week be altered through each successive curse. Note also that each curse was in response to humanity's sin. The first of the curses was pronounced right after Adam and Eve first sinned, and can be found in Genesis 3:17-19: "And unto Adam he said, Because thou hast hearkened unto the voice of thy wife, and hast eaten of the tree, of which I commanded thee, saying, Thou shall not eat of it: cursed is the ground for thy sake; in sorrow shalt thou eat of it all the days of thy life; thorns also and thistles shall it bring forth to thee; and thou shalt eat the herb of the field; in the sweat of thy face shalt thou eat bread, till thou return unto the ground."

There are a couple fascinating things about this curse. First, note that the vegetarian diet of human beings was confirmed in line with the special dietary instruction given in Genesis 1:29, 30. Sinful humans were not yet given leave by God to eat animals. Second, note what God says the earth will bring forth as human beings labor in the field: "thorns also and thistles shall it bring forth to thee." Christians popularly believe that God created the rose and Satan added the thorn. From this verse we know that thorns and thistles were part of a curse placed on the earth by God that ultimately worked to suppress food production, at least to the degree than humans had to work hard ("in the sweat of thy face shalt thou eat bread") to get food.

As to the second curse, we are told that "in the days of Noah a double curse was resting upon the earth in consequence of Adam's transgression and of the murder committed by Cain."[11] The second curse that changed humanity's relationship with nature, and may have changed the actual systems that make up the functioning of the earth, is to be found in Genesis 4:11, 12: "And now art thou cursed from the earth, which hath opened her mouth to receive thy brother's blood from thy hand. When thou tillest the ground, it shall not henceforth yield unto thee her strength."[12] It is interesting to note that each of the three curses that lay upon the earth (we will get to the third curse momentarily) as a result of sin have impacted humanity's ability to gain food from the earth. In later chapters we will see how modern human beings have briefly overcome these three curses, to disastrous effect, and we will also see that the combined effects of sin upon the global systems of the earth act to suppress food production upon the earth, with dire results in our current age.

We read that when God decided to send a flood upon the earth in response to humanity's wickedness, He would also "destroy the things with which He had delighted to bless them; He would sweep away the beasts of the field, and the vegetation which furnished such an abundant supply of food, and would transform the fair earth into one vast scene of desolation and ruin."[13] We also read that "the entire surface of the earth was changed at the Flood. A third dreadful curse rested upon it in consequence of sin."[14] We do know that the original design of earth's food production systems, starting with the divinely created irrigation system and extending to the atmospheric and oceanic systems,[15] was drastically changed through this third curse. And we do know that humanity originally did not have to exert effort to gain food, because the earth was so amazingly bountiful that humans had but to extend their hands and take of the food that was in constant supply. The combined result of sin and God's three curses upon the earth changed that forever.

Three curses would seem sufficient, but there is one more curse that God placed upon the earth, and as we study the effects of sin upon God's global systems, we must also keep this curse particularly in mind and try to weigh its effects, as it may be responsible for some of the phenomena that we see reflected in nature. We read that "all nature is confused; for God forbade the earth to carry out the purpose He had originally designed for it. Let there be no peace to the wicked, saith the Lord. The curse of God is upon all creation. Every year it makes itself more decidedly felt."[16] We are also told that this particular curse is upon everything, and is related to the intensity of transgression. "The Lord's curse is upon the earth, upon man, upon beast, upon the fish in the sea, and as transgression becomes almost universal the curse will be permitted to become as broad and as deep as the transgression."[17]

Finally, there is this interesting passage in the book *Education:* "Although the earth was blighted with the curse, nature was still to be man's lesson book. It could not now represent goodness only; for evil was everywhere present, marring earth and sea and air with its defiling touch. . . . Thus not only the life of man, but the nature of the beasts, the trees of the forest, the grass of the field, the very air he breathed, all told the sad lesson of the knowledge of evil."[18] The curse of sin pervaded everything in the earth.

The combined weight of sin, the mismanagement and maliciousness of Satan, and the curses of God now rest upon the earth, causing the perfect and perfectly balanced systems that underpinned God's original creation to continuously degrade and decay through the ages. It is useful for every Christian that is concerned about preparing the way for the Lord in these fleeting last days to be aware of this continuing decay, because by being aware, we will hear the footsteps of our approaching Lord in each headline about weather catastrophe, or famine, or increases in food prices or energy prices, or news of environmental decay or climate change. And if we understand this decay, we will be better able to explain to a wondering world that the disasters and decay in the world are a result of sin, and that Christ is coming. Let me say that again: The state of the world shouts that Christ is coming! This, paired with the message that Satan is the author of suffering and disaster, is a powerful message to accompany the teaching of the gospel.

In coming chapters we will look at some of earth's individual systems that are rapidly decaying and will get a better feel for how the impact of this decay matches the expected state of the earth at the start of the time of trouble. But first we will explore how the Lord feels about the decay and destruction of the earth He created and how the earth itself, and not just humanity, is part of the great controversy and the plan of redemption and restoration.

---

[1] See Ellen G. White, *The Great Controversy* (Mountain View, Calif.: Pacific Press Pub. Assn., 1911), pp. 494, 495. See also Isa. 14:12-14.

[2] E. G. White, *Patriarchs and Prophets*, p. 67.

[3] *Ibid.*, p. 62.

[4] *Ibid.*, p. 52.

[5] *Ibid.*, pp. 67, 68. (Italics supplied.)

[6] *Ibid.*, p. 57.

[7] *Ibid.*

[8] *Ibid.*, p. 61.

[9] We don't know what the "terms of engagement" are in the war between Christ and Satan. We know that God allows Satan enough control over the earth so that he cannot credibly claim that God never allowed him to demonstrate fully his administration of the earth and the net effects of that administration. Proving the failure of Satan's way is, after all, the point of the great controversy. As to the degree of control, Ellen White says in *The Great Controversy* that God will decrease His restrictions on Satan in the last days. So we know Satan will exercise more power, and we must assume he will substantially control all of nature in the final days.

[10] E. G. White, *Patriarchs and Prophets*, p. 90.

[11] *Ibid.*

[12] It can be argued that this curse was only on Cain, but in *Patriarchs and Prophets* (p. 107) Ellen White writes of a third curse resting on the earth after the Flood, so we know that Cain's curse affected the entire earth.

[13] E. G. White, *Patriarchs and Prophets*, p. 92.

[14] *Ibid.*, p. 107.

[15] It is not possible to go from a nonraining atmospheric system to a raining atmospheric system without fundamentally changing the system itself. In addition, if the atmospheric system were changed, the oceanic system (which exchanges gases with the atmosphere) would also be affected, and it is extremely likely that other global systems and subsystems would be strongly impacted or eliminated by such a fundamental change.

[16] *The Seventh-day Adventist Bible Commentary* (Washington, D.C.: Review and Herald Pub. Assn., 1953), Ellen G. White Comments, vol. 1, p. 1085.

[17] *Ibid.*

[18] Ellen G. White, *Education* (Mountain View, Calif.: Pacific Press Pub. Assn., 1903), pp. 26, 27.

Chapter 3:

For God So Loved
the _____

"For God sent not his Son into the world to condemn the world; but that the world through him might be saved." John 3:17.

L et's take a look at how God regards the world He created, and for the sake of clarity let's stipulate to a couple key points to make sure we start at the same place:

1. Christ our Redeemer died to save us, whom He created in His image. To be saved, we must accept this sacrifice and claim Christ as our Savior.

2. The earth that we live on will be completely destroyed and will be re-created without any stain of sin.

The above two points are pretty much standard thinking among Christians and are, of course, derived directly from Scripture. However, the two above points are not often connected in Christian thinking. Nor, for that matter, do Adventist Christians give voice to the logical conclusion to the line of thinking inherent in the two points, which is to say: "Therefore, the earth does not really matter; only people do." This conclusion, which seems to be widespread but unconsciously arrived at, would be perfectly reasonable if not for the fact that both Scripture and inspired writings do not support this conclusion and, if anything, are at odds with it. The disparity between our unconscious conclusions and the express words of Scripture present an opportunity for study and reflection, and in that spirit let's take a look at God's attitude toward the earth itself, and how the physical earth fits into the culminating war between Christ and Satan.

To frame the study, let's extrapolate some points of the nature of the Creator based on the fact that we were created in His image. Occasionally when humans engage in some act of creation, say a painting or written work or new musical work, and achieve the design they were intending to create, then it is normal to take satisfaction and pride in their work and to assert ownership of it. In cultures around the world, this is what humanity does. The question is whether God also did these things in regard to the creation of the earth. For clarity, let's form two distinct questions:

1. Did God take satisfaction and pride in the created earth?

2. Did (and does) God assert ownership of His created earth?

With the questions properly framed, let's go to Scripture and inspired writings to see what we can find. Perhaps the best place to start in regard to the first question is the account of Creation in Genesis 1. Recall that after each major system was created, God reviewed His work and saw that it was good, and that He reviewed the whole of creation when He was done and declared it *very* good. On the basis of God's own assessment of His work, it seems reasonable to understand that God took satisfaction and pride in the created earth, especially since this makes it clear that God met His objectives in His creation acts. This view is buttressed by the establishment of the Sabbath as a perpetual day of rest and a perpetual memorial to the acts of creating the earth (see Gen. 2:3). Logic strongly suggests that if God were not pleased with His creation, He would not have eternally memorialized His acts of creation with the Sabbath. Other writings expressly support the view of a Creator pleased with His work: "God looked with satisfaction upon the work of His hands. All was perfect, worthy of its divine Author, and He rested, not as one weary, but as well pleased with the fruits of His wisdom and goodness and the manifestations of His glory."[1] A passage in *The Desire of Ages* reinforces the point: "In the beginning the Father and the Son had rested upon the Sabbath after Their work of creation. When 'the heavens and the earth were finished, and all the host of them' (Gen. 2:1), the Creator and all heavenly beings rejoiced in contemplation of the glorious scene. 'The morning stars sang together, and all the sons of God shouted for joy.' Job 38:7."[2] Here we have not just God taking satisfaction, but all of the universe rejoicing. With these references we can consider the question settled—God did indeed take satisfaction and pride in His created earth.

The second question—"Did (and does) God assert ownership of His created earth?"—is just as straightforward in its answer. God clearly asserts ownership while also clearly acknowledging that Satan has temporarily usurped dominion over the world. In 1 Chronicles 29:11 we read the words of David: "Thine, O Lord, is the greatness, and the power, and the glory, and the victory, and the majesty: for all that is in the heaven and in the earth is thine; thine is the kingdom, O Lord, and thou art exalted as head above all." God has not given up His claim on the world, but the Bible does explain how Satan used the Fall of humanity to usurp dominion over the earth. In 2 Peter 2:19 we read that "of whom a humanity is overcome, of the same is he brought in bondage." In other words, humanity, who had dominion over the world, was conquered by Satan, and thus Satan became "the god of this world" (2 Cor. 4:4) because humanity is in bondage to Satan. In *Patriarchs and Prophets* we also read this express point, and the same scripture is used to underscore the point.[3]

As further regards an assertion of ownership and as an expression of His displeasure over what Satan and sin have done to the world, God has this to say in Revelation 11:18: "And the nations were angry, and thy wrath is come, and the time of the dead, that they should be judged, and that thou shouldest give reward unto thy servants the prophets, and to the saints, and them that fear thy name, small and great; *and shouldest destroy them which destroy the earth.*" This text takes on a distinct meaning when we consider what sin has done to God's creation. From this verse we know that God is very displeased with the corruption of His creation.

To understand more fully God's attitude toward the earth, we need look no further than the most familiar, most memorized text in the Christian world—John 3:16. Familiar it may be, but few, apparently, have meditated on its broader meaning. In John 3:16 we read: "For God so loved the world, that he gave his only begotten Son, that whosoever believeth in him should not perish, but have everlasting life." This is a compound verse, and in it a distinction is made between the world and the people in it. God gave His Son to redeem the world and all that is in it from Satan, but the people of the earth—the free moral agents—must believe on Him in order to be redeemed. And so it is that humans must accept Christ in order to be saved, and in order for the world to be redeemed from Satan. We humans play a key role in the restoration of everything that Satan took, which is as it should be since it was by our sin in the Garden of Eden 6,000 years ago that Satan was given dominion over the earth.

One of my favorite writers gives additional insights into God's claim as owner of the world and Satan's usurpation, and pretty much wraps up in one sentence one of the major points of this chapter and this book: "Not only man but the earth had by sin come under the power of the wicked one, and was to be restored by the plan of redemption."[4] So there we have it in as succinct a statement as it is possible to have: The war between Christ and Satan is not just about humanity—it is about everything that Satan took. God wants it all back—the earth, the people, and all that He made during Creation week—and He was willing to sacrifice His Son to redeem creation and blot sin from the universe once and for all.

Thus it is that the earth is not just the stage on which the great controversy is played out, but it is also part of what is at stake while at the same time serving as a prime example of the effect of sin—death. The earth itself bears the mark of sin in its degraded and increasingly unstable systems, and the arc of cause and effect between original sin and the current state of the earth is growing clearer and clearer in our present age.

The decay of the earth's systems has reached the point where the condition of the earth now begins to approach the descriptions of the earth given in Mathew and inspired writings in reference to the state of things just before the coming of Christ. For dedicated Christians, the state of the systems of the earth in our age should be a signal of paramount importance, since each sign of decay and degradation means that the time of trouble is drawing closer and closer, which in turn means that the time left in which to labor for our Lord grows shorter every day.

But more than providing additional urgency, the study of the decay of earth's systems can provide sharp focus by illuminating and adding specific knowledge to the warnings and counsel of Scripture and inspired writings and by providing knowledge that will help Christians to tailor ministries specific to the events and impacts that are unfolding around us.

The balance of this book will be spent looking at specific systems that God created, the signs of degradation and instability in these systems, and how the decay of these systems is ushering in the time of trouble in our current age. Following chapters will look at the effects of sin in our food production systems, in our climate systems, in our oceanic and freshwater systems, and how the effects of decay and instability in these systems interact with the global systems created by humanity—energy and financial systems. We will also look at how the societies of earth will be dramatically impacted and suffer horribly in our age as the foundations upon which they are built rapidly erode. Finally, we will look at what we can do to prepare ourselves, our families, and our neighbors for what is coming, and how we can create or redirect ministries so that the services we offer are responsive to what is coming upon the earth.

As we study the systems of the earth, we will be looking in each at the effects of sin and the degree to which global systems are on the verge of fulfilling the prophecies regarding the state of the world just prior to the time of trouble. With this in mind, it is fitting to revisit Matthew 24, where the disciples approached Christ privately and asked Him to explain the signs of His coming and of the end of the world. In verses 6-8 Christ gave a succinct answer: "And ye shall hear of wars and rumours of wars: see that ye be not troubled: for all these things must come to pass, but the end is not yet. For nation shall rise against nation, and kingdom against kingdom: and there shall be famines, and pestilences, and earthquakes, in divers[5] places. All these are the beginning of sorrows."

With these verses in mind, let's look at the systems of the earth and see how close we are to meeting Christ's description of the earth at the time of the "beginning of sorrows," or at the beginning of the time of trouble.

---

[1] E. G. White, *Patriarchs and Prophets,* p. 47.

[2] Ellen G. White, *The Desire of Ages* (Mountain View, Calif.: Pacific Press Pub. Assn., 1898), p. 769.

[3] E. G. White, *Patriarchs and Prophets,* p. 67.

[4] *Ibid.,* p. 67.

[5] I love the King James Version, as it is so eloquent, but occasionally changes in our language make it difficult to access. In this case, "divers" does not mean the bottoms of lakes and seas where divers go—it means "diverse," or that earthquakes shall occur in many places.

Chapter 4:

The Decay of Our Global Food Production System

"The present is a time of overwhelming interest to all living. Rulers and statesmen, men who occupy positions of trust and authority, thinking men and women of all classes, have their attention fixed upon the events taking place about us. They are watching the relations that exist among the nations. They observe the intensity that is taking possession of every earthly element, and they recognize that something great and decisive is about to take place—that the world is on the verge of a stupendous crisis."
—Ellen G. White, Prophets and Kings, p. 537.

When I was a boy growing up in southern California, my parents had a fair-sized garden in our backyard where we grew tomatoes, spinach, carrots, squash, beans, beets, and, most important to a young boy, watermelons and cantaloupes. Every spring my dad would head to the backyard and turn the entire garden area over with a spade. I loved following him around and poking at the fragrant, freshly turned earth. More often than not, there were pink earthworms in the soil, and sometimes there were cutworms or other damaging pests. My dad taught me to leave the earthworms alone because they were "good for the garden," but to gather up the pests and put them in a bucket.

After all the soil was turned over, my dad would take a rake to the garden area and rake the fine, loamy soil until it was as flat and smooth as a tabletop. Then he would set up stakes at each end of the garden, stretch string between them, and plant carefully distanced seeds along the string. Sometimes my dad would let me plant seeds at the ends of rows. I of course wanted to grow a large amount of vegetables, so naturally I would plant a large amount of seeds in the area he gave me. Many was the time he told me that only a few seeds were needed in a small area in order to harvest a good crop. When the seeds sprouted, my dad would take me out to the garden and direct my reluctant thinning of the seedlings that had come up where I had planted.

As spring turned into summer, the seeds that my dad had planted turned into large, healthy plants that filled every part of the carefully laid-out, irrigated garden. In summer and fall our table was regularly filled with the healthy abundance of our garden, and my mom canned the excess or gave it to others who could use it.

Growing up this way, I thought that producing a large harvest was a fairly easy, straightforward matter; except for the arduous task of weeding, you simply planted seeds and then harvested your crop a few months later. It was not until my wife and I moved to La Vida Mission, a Navajo Indian ministry in a remote part of New Mexico, that I began to get an appreciation for how difficult it can be to grow a garden. The rich, loamy, bottomland soil that I had grown up with in California was strikingly different from the sand-and-clay alkali soil at La Vida. Still, I tried to turn over some Navajo reservation earth for a small garden. It was real work—what wasn't hard clay was stones and sand. I knew the poor soil needed some kind of help, so I got some sheep manure from a Navajo herder a few miles away and spaded that into the soil. I planted a few squash, which are one of the few crops known to tolerate the soil in that area, and watched my garden expectantly, watering it with scarce well water when rain was insufficient, as was the case most of the time.

The squash plants were small, sickly, and never set a blossom (despite my attempts to help through hand pollination). The squash venture was, to put it succinctly, a total failure. This experience marked the start of my education on the difference between "productive" and "unproductive" soils and the relative lack of "productive" soil around the world. Over the years I continued to learn about soil and food production, culminating with my learning that, on a global basis, a fair amount of our soil is being or has been degraded at the same time that decay in our other global systems is pressing agricultural yields down. We are in a situation in which a significant amount of the fertile soil on the earth, such as I experienced when I was a child, is headed toward a productivity level such as I experienced in the badlands of the Navajo nation. It is difficult to overstate how important this is on a global basis, and yet it is just one of many significant problems now facing our global food production system.

One of the first things you notice when you start looking in detail at the degradation of the earth's systems is that they are interlinked and interdependent—like the gold pocket watch discussed in chapter 1—and that it is very difficult to try to view them in isolation, because they certainly do not function in that manner. Our global food production system, which is highly integrated with other global systems yet is not itself a distinct "system" as are the others we will look at (such as fresh water and oceans and climate), is distinguished by two things: first, it is the "system" that is perhaps most measurably impacted by any negative change in any of the global systems that God created, and is also the system that is most integrated with human-made global systems, such as our global energy (oil) production and distribution system and our global financial system; changes or shocks in these human-made global systems also affect food production negatively. The second thing that distinguishes the system is that it has the weight not just of sin on it, but also the weight of three separate curses from God. As a result, our global food production system is the most precarious and at-risk of all systems. In terms of measuring global system decay, it is the canary in the coal mine.

Recall from previous chapters that God first cursed the earth and limited its productive capacity after Adam and Eve sinned. Let's review the curses briefly. In Genesis 3:17-19 we read the first curse: "Cursed is the ground for thy sake; in sorrow shalt thou eat of it all the days of thy life. Thorns also and thistles shall it bring forth to thee; and thou shalt eat the herb of the field; in the sweat of thy face shalt thou eat bread, till thou return unto the ground." Interestingly, the first curse is delivered "for thy sake" and forced humans to work hard for the food they would get from the earth, instead of living a life of leisure and idleness. We will revisit this point at the end of this chapter. The second curse followed soon thereafter, when Cain slew Abel. We read of this curse in Genesis 4:11, 12: "And now art thou cursed from the earth, which hath opened her mouth to receive thy brother's blood from thy hand; when thou tillest the ground, it shall not henceforth yield unto thee her strength." The third curse weighing on the earth is the Flood, which forever altered the way water, nutrients, and elements are cycled through the atmosphere and soil, causing yet more decay in its productive capacity.

Our global food production system has an array of problems, and we will cover only a few of them here. That said, perhaps the biggest problem for the food production system is outside of the food production system itself, and that is the burgeoning world population that the dwindling amount of productive land must feed. On page 41 is a simple chart showing the growth in world population from the time of Christ, when some 200 million people are estimated to have lived on the earth, until today, when we have more than 7 billion people alive and competing for resources. A glance at the chart is all that is required to draw the conclusion that the situation is not sustainable.

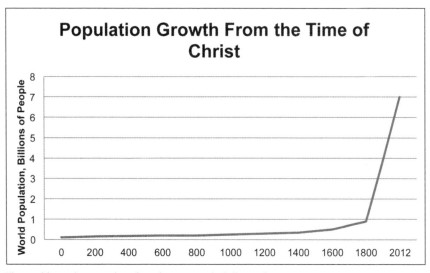

Population Growth From the Time of Christ

The world population is thought to have topped 7 billion in late 2011.

The post-Flood earth has never had so many mouths to feed—and it has never produced as much food as it does at this writing. Interestingly, it is the abundance and availability of food over the past 150 years that has helped facilitate such a population burst, and it is the availability of oil that has helped bring about the availability of food. But we are getting ahead of ourselves. Let's list some of the major problems with our global food production system[1] and then look at each of them in some detail:

1. Oil is food.
2. Food is oil.
3. Degradation of soil and desertification
4. Erosion of soil
5. Water shortages
6. Climate creep
7. Resurgence of pestilences and disease
8. Phosphorous depletion

Oil Is Food: Throughout much of the world, our food production is industrialized. This means that, yes, our store-bought tomatoes taste like cardboard, but far more important, it means that agriculture is carried out on a vast scale that is mechanized and energy-intensive, with oil as the most important input. Industrial agriculture requires chemicals, pesticides, and fertilizers, almost all of which are made out of oil or natural gas. It also means that our crops require constant tending by heavy equipment that in turn requires fuel made from oil. Harvesting requires more heavy equipment and, of course, more oil, while processing, packaging, and transportation are all energy-intensive activities that require significant additional amounts of energy, quite often in the form of oil. Put simply, we convert oil and sunlight into food. And we are not very efficient about it. In fact, we use about nine calories of oil and one calorie of sunlight to make one calorie of food.[2] Oil has roughly 31,000 calories per gallon,[3] so we actually manage to make about 3,400 calories of food (on our plate) with that oil. Americans have available about 3,800 calories per person per day,[4] so we need about 1.1 gallons of oil per day to feed each of us, or more than 2,000 gallons of oil per year to feed a family of five. This amounts to almost 50 barrels of oil, which means, with oil at $100 a barrel (as it is as of this writing), that the oil "content" in our food costs a family of five almost $5,000 per year. And of course, any increase in the price of oil will mean an increase in food prices.

This is a sobering fact for a number of reasons. First, and rather obviously, it means that the cost of oil now makes up a significant part of the cost of food, and if oil increases at all in price (and it has increased significantly in the five years prior to the writing of this book), then the cost of food must also increase. This is very bad news for the roughly 3.5 billion people on earth who live on $2.50 a day or less[5] (for reference, 3.5 billion people is more than 10 times as many people as now live in the United States and is half the people on the planet). Think about that for a moment: Half the people on the earth cannot afford the cost of the oil that is a part of the food they must eat every day. Second, our reliance on oil in food production means that any decrease or disruption in the availability of oil turns into a decrease or disruption in the availability of food, which again will impact the poor of the world the most. Third, the heavy use of oil in industrial agriculture also means that agriculture is a significant producer of greenhouse gases that disrupt the atmospheric system God created. As we will discuss later, excess greenhouse gases create effects that ripple through the earth's systems, including agriculture. The abundance of oil over the past 150 years (and its relative cheapness for much of that period) has meant that societies around the world have been able to harness fossil fuels to produce a great amount of relatively inexpensive food. The almost uniform response to this abundance of food (and other factors such as a decrease in disease) in countries around the world has been for people to have more children. During the period that oil has been available—the past 150 years or so—the global population has exploded. Now, the roughly 7 billion people on the face of the earth are wholly dependent on industrial agriculture, which in turn is wholly dependent on vast quantities of "cheap" oil being delivered on an uninterrupted basis. Of the 7 billion people on earth, roughly 15 percent already go hungry regularly, are malnourished, and face grave peril or death if their food supply is disrupted. A much larger number is at risk of hunger and malnourishment if their food supply is disrupted. This, obviously, is a massive problem as we shall see when we explore oil and humanity's energy systems (and the fragility of complex societies) further in chapter 8.

Food Is Oil: Some discussion in the media has been given over the past five years to the fact that "biofuel" production, especially in the United States, is reliant upon farmland and is displacing food in order to create fuels, thus decreasing the amount of food available and increasing its cost. Others have asserted, somewhat plausibly, that the production of such "biofuels" is very inefficient and is so reliant upon oil (see above) that they result in little or no greenhouse gas reduction, and perhaps even add to the problem.[6] While the biofuel sector is trying to move to "cellulosic" fuels,[7] for the present the use of food products, particularly corn, to produce biofuels continues, and if fuel shortages were to strike on a global or regional basis, governments would have to make very difficult decisions on how much food will be produced and how much fuel will be produced from a limited supply of cropland.

Degradation of Soil: In the late 1950s and especially in the early 1960s the world was facing a food crisis. The "huge" population (the population was then less than half what it is now) had grown to the point at which the production from the available farmland was not keeping up with demand, and famine was on the horizon.[8] The leaders and scientists of the time launched something called the "Green Revolution," which involved a two-part plan for solving the problem: The first part involved breeding plants to get better production from the same acreage. The second part involved using fertilizers and pesticides to, again, get better production from the same acreage. The plan worked brilliantly. Production rates soared the world over. But there were unforeseen consequences. For one thing, world population continued to grow at a dizzying pace, mostly because plenty of food was available[9] because of the success of the Green Revolution. The population boom enabled by the Green Revolution is an extremely serious thing, but there were other consequences that were less obvious and perhaps more dangerous. Foremost among these is the fact that the Green Revolution was the springboard for the global industrialization of food production and was the catalyst for heavy global use of fertilizers and pesticides, particularly in impoverished nations.

Ironically, one of the most damaging of the unforeseen consequences of the Green Revolution has been the degradation of productive soil in many parts of the world as a result of using fertilizers and pesticides. Almost all fertilizers and pesticides are produced from fossil fuels—oil and natural gas. When fertilizers are used on farmland, part of the fertilizer is absorbed by the plant, part is washed away by irrigation or rain (more on the serious consequences of this in a later chapter), and part remains in the soil as salts. Each year, as excess additional fertilizers are added, more and more salts build up in the soil, and eventually start killing the microorganisms that make soil productive. In parts of the world where fertilizers and pesticides are chronically overapplied, productive soil has been turned into sterile dirt that does not support plant growth and that erodes very easily. An example of this is China,[10] but overapplication of fertilizers is widespread. Not much of the world's agricultural land has been rendered completely useless, but almost all of it is reliant on fertilizers and pesticides to maintain the production levels required to feed the world's population. Any disruption in fertilizer and pesticide supply means farmers will not be able to produce the food the world needs, and any continued use of significant amounts of fertilizers and pesticides means that the limited amount of productive soil available on the earth will significantly degrade. To give a feel for the extent of the current soil degradation, since 1945 more than 4.5 billion acres of soil have been degraded, and this figure breaks down into categories of degradation, with 38 percent being lightly degraded, 46 percent being moderately degraded, 15 percent being severely degraded, and the balance (1 percent) being so degraded that no effort will reclaim it.[11]

A further factor in soil degradation is mineral depletion. Crops (particularly corn) are hungry for minerals, and when the crops are harvested (as opposed to decomposing in the field), the minerals are not returned to the soil. Fertilizers generally add only a few of the minerals that are taken from the soil, meaning that aggressive continued cropping leads to mineral depletion, which further suppresses production. The logical solution to this problem is to move on to new land, which is exactly what slash–and–burn agriculture does—people in developing countries cut down forests, burn the area to clear out wood and vegetation and release the minerals, crop the area intensively until the soil is depleted (often only a year), then move again and repeat the cycle. The soil left behind can no longer support forests or crops, is considered severely degraded, and is subject to significant erosion.

Erosion of Soil and Desertification: If we count all the land that has, since 1945, been degraded through soil depletion, desertification, and deforestation, the total is more than 11 billion acres, or about 45 percent of the earth's vegetated surface.[12] Each year more than 20 million acres of land that was once forest—and then farmland—are abandoned because of severe degradation, and does not revert to forestland.[13] This land must be replaced (often through slash-and-burn techniques). In addition, about half again as much new land must be brought into agricultural production to feed our growing population. The problem of finding new land to farm is exacerbated by the fact that we have subjected so much of our current cropland to erosion and are consuming it at unsustainable rates. When more and more cropland around the world reaches a severely degraded state, there is no apparent place to turn for new cropland. A prime example of this is China. In my time in China I witnessed firsthand the advanced state of degradation of farmland resulting from the overapplication of fertilizers and pesticides, as well as mineral depletion and erosion. In the 1999 to 2001 period I witnessed entire villages being moved—houses and all—onto degraded and depleted land so the soil beneath the former villages could be farmed. China, with its intense pressure to produce food, is ahead of the rest of the world in terms of soil degradation and erosion, but not by much. China is, at the same time, an example of a state willing to invest enormous amounts to attempt to recover its farmland; such an enormous societal investment increases the cost of food, and only relatively wealthy societies can attempt such an effort.

Erosion of soil happens primarily in two different ways: water action and wind action. In both instances soil is carried away, and in both instances erosion is made possible by a loss of cover (the plants that cover the soil and whose roots hold the soil in place) and by such poor farming practices as the overapplication of fertilizers and pesticides previously discussed, but also extending to poor tilling practices. Scientists say that the formation of topsoil occurs at a rate of .25 to 1.5 millimeters per year, depending on climate and other factors.[14] This means that, on average, land that is not used for agricultural purposes (where vegetation decays and builds up each year) forms top soil at a rate of a little more than four tons per acre per year. Unfortunately, where agriculture is practiced, topsoil is eroding at a rate that sometimes exceeds 40 times its natural formation rate.[15] There is a word for earth that has been depleted of topsoil: sand.

Wind-eroded topsoil is a particular problem for China, where as many as 100 million people may be "erosion refugees" as their land becomes unusable in 30 years or less.[16] The loss of topsoil decreases productivity by up to 65 percent,[17] and the continued erosion of the soil steadily decreases the food production capacity of the earth. The earth is only so big, and there are no realistic options for bringing a sufficient amount of new lands into production. The earth as a whole is, it seems, moving very rapidly in the direction of China and its desperate measures to maintain food production.

Water Shortages: We will review the global freshwater system in detail in a later chapter, and thus we will deal only briefly with this complex subject here. It is perhaps best to start again in China, if for no other reason than the author's personal familiarity with the situation there. In China, water tables are dropping precipitously. In some areas farmers that installed shallow wells and pumps 20 years ago in order to irrigate their land have now dug their wells up to 400 feet deep, and still the water table is dropping away from them. The water shortages throughout much of China are so severe that water is in deficit in half of the largest cities.[18] In India multinational companies have installed numerous large wells and massive pumps in order to extract water for bottling and sale.[19] At the same time, overpumping by almost everyone with a well is lowering aquifers across that country.[20] The net effect is that the farmers who once relied on their own wells to irrigate their land now find their wells have gone dry and must buy water for even their personal use from the companies that are bottling it. Even more worrisome is the fact that water problems in India are not just an Indian problem but potentially a global crisis, since there are already water disputes with Pakistan and China that could lead to war.[21] Throughout the world it takes an average of about 1,000 tons of water to produce one ton of grain.[22] The aquifers in many countries are seriously depleted, and yet the people still demand food, and in particular they demand grain.

In the United States the situation is no less dire. Throughout the United States, aquifers are being pumped at far beyond their recharge rate. This is true in the Southwest, in the South, and throughout the Midwest, where the Ogallala aquifer (which underlies some of the most productive cropland in the country) is being pumped at far beyond its recharge rate and is expected to become depleted within the next 20 years or so.[23] Loss of the Ogallala aquifer will dramatically impact food production (and cost and availability) in the United States and beyond. In short, the supply of water used for global agriculture is decreasing rapidly, with dire consequences expected, and with no effective solution currently envisioned.

Climate Creep: As if the "perfect storm" of threats to industrial agriculture were not enough, there is also the threat of climate creep, which is probably one of the most significant threats, even though it is not often specifically addressed in popular media. To understand the threat of climate creep, it is first necessary to understand that our modern agricultural practices were deliberately located. That is to say, farmers chose the best land they could find for the conditions that prevailed at the time, and proceeded to grow crops there. With climate creep, land that was once ideally situated and well watered is now subject to a host of different conditions, and in general these new conditions serve only to suppress agricultural production.

Resurgence of Pestilences and Disease: Under climate creep, formerly ideal farmland may be wetter or drier, and is subject to a host of new challenges including molds, fungus, and other plant and animal diseases and pestilences.[24] Of these, it is pestilences that are perhaps the most worrisome, especially since we are already losing the war on pests and weeds. From the 1940s to the 1990s, crop losses to pests and weeds *increased* despite a tenfold increase in both the amount and toxicity of pesticides and a hundredfold increase in herbicide use.[25] We cannot afford more losses on this front, but with climate creep comes threats to our agriculture that we simply have not had to deal with in the past. The coming chapter on decay in our global climate system will look at this aspect of the threat in more detail.

Globally, farmers are in a bind because, simply put, they cannot move their land, and yet the conditions in which they farm have begun to change, bringing new challenges that farmers are not prepared or equipped to deal with, and new conditions that in all likelihood are nowhere near ideal for growing the crops that were previously grown. All of these factors make agricultural production far less stable and predictable, and inevitably increase the cost of food.

Phosphorus Depletion: The depletion of phosphorus reserves worldwide is one of the gravest resource shortages of our time, and yet it is likely you have not heard of it. Use of phosphorus surged with the Green Revolution; the critically important mineral is used to substantially boost crop yields, and without it harvests plummet. Without phosphorus, the Green Revolution would have been a flop. Current global use of phosphorus exceeds 17 million metric tons and is growing at 3 percent per year.[26] Current reserves of phosphorus, of which 90 percent is concentrated in just five countries,[27] are expected to be exhausted by as early as 2040.[28] Since the resource is critically important, the looming shortage is expected to kick off an intense international resource competition. Prices of the mineral have already surged, increasing costs to farmers and ultimately increasing food costs. While adaptation and resource recovery (such as recovering phosphates from sewage water) may help extend the resource by a number of years, such measures do not seem capable of significantly avoiding resource exhaustion. The price of phosphate rock surged from about $60 per metric ton in 2005 to almost $350 per metric ton in 2008 before falling in the global recession and ending 2011 at about $200 per metric ton.[29] The chart below shows the projected availability of phosphate rock through 2050, and reflects the fact that production peaked in about 1990 and has been in a downward trend since then.

World Rock Phosphate Production

Source: www.energybulletin.net

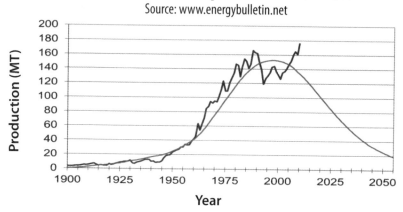

Adding up everything we have just covered, the situation looks remarkably grim—and this is without factoring in the influence of the decay of other systems, all of which negatively impact food production. And yet, humanity was created in the image of God, and even in their decayed state, human beings are pretty smart and can achieve much, so the question is whether or not human beings can fix the mess that the global food production system is in. It seems that the answer lies in the fact that humans have not, at least since the Tower of Babel, cooperated on a "global" scale to achieve a coordinated undertaking. Even if they were to cooperate in such a way, humans at their best seem capable of focusing their efforts on overcoming one or two large obstacles at a time. There is no precedent to suggest that human beings are capable of tackling, on a global basis, the "perfect storm" of hurdles faced by our global food production system. The obvious conclusion is that, with the threats faced by the world, there seems to be no viable way to continue feeding the teeming masses that populate the earth.

But even without an obvious solution humanity must attempt to solve the problems in the sector, because our global societies are built, in the final analysis, on industrial agriculture. The wealth, power, and technology of the societies around the world ultimately depend on the ability of humanity to keep those same societies from starving. And some of the adaptations and desperate solutions that sinful human beings will likely adopt are not likely to be peaceful. Given the nature of humanity and the state of nations at this point in history, stronger nations are almost sure to use their might to take resources and territory from weaker nations. Doing so will make perfect political sense, since actual solutions are so difficult, so unsure, so expensive, and so time-consuming that they cannot be pursued without also simultaneously pursuing a "Plan B" that takes needed resources from weaker nations. The net effect of denial in regard to these problems is that, when they are finally acknowledged, the opportunities for viable solutions are so far past that only desperate and aggressive measures are left as options. Ultimately, the use of force will allow nations to compensate for lack of planning and a lack of sustainability in industry, agriculture, transportation, etc.

Within this context, lets revisit Matthew 24:6-8 and see what Christ had to say about the state of the world just before the time of trouble: "And ye shall hear of wars and rumours of wars: see that ye be not troubled: for these things must come to pass, but the end is not yet. For nation shall rise against nation, and kingdom against kingdom: and there shall be famines, and pestilences, and earthquakes, in divers places. All these are the beginning of sorrows." With the exception of earthquakes, all of the conditions mentioned above can be fulfilled by the challenges posed by the breakdown of our global food production system. Think on that for a moment—we have only begun to look at the challenges that face the societies of humanity because of sin-induced global system decay, and already we can see some of the forces at work that will bring the world into a state as described by Christ in Matthew 24:6-8.

We started our review of the systems that are decayed and dying by looking at food production, and we did so because food production was singled out and specifically cursed by God three times. At the beginning of the chapter it was also promised that we would revisit the first of those curses and look at it a little more deeply. The first curse is found in Genesis 3:17-19: "Cursed is the ground for thy sake; in sorrow shalt thou eat of it all the days of thy life; thorns also and thistles shall it bring forth to thee; and thou shalt eat the herb of the field; in the sweat of thy face shalt thou eat bread, till thou return unto the ground." There are two striking things about this curse—first, that it was given "for thy sake," and second, that the curse appears to be in suspension for most of the people around the world as a result of industrial agriculture. These two points are related, so let's look at them closely.

Some interesting things have been written that touch on these verses. We read that "the land has felt the curse, more and more heavily"[30] and that "the sin of man has brought the sure result—decay, deformity, and death. Today the whole world is tainted, corrupted, stricken with mortal disease. The earth groaneth under the continual transgression of the inhabitants thereof."[31] We are also told that we are to work in the soil with our hands for our own good, growing our own food, as it will bring us closer to God and His nature.[32] Aside from the benefits to be had from gaining our food by hard labor, there is also the fact that food that is hard to come by more or less automatically results in smaller families. In other words, one of the effects of an unproductive earth is population control—part of the "for thy sake" in God's curse on the ground seems to be so that humanity's population will not explode.

Look again at the population chart at the beginning of this chapter— it was not until man gained an "energy subsidy" though the use of fossil fuels (first coal, then oil and gas) that population growth soared. In other words, the mechanization, irrigation, and fertilizers and pesticides that have come into almost universal use during the oil age have given humans far more energy (or horsepower if you prefer) than would have otherwise been available, thus allowing them to escape, for a time, the curse of Adam. The results of this energy subsidy and the temporary escape from the curse of Adam seem to be uniformly negative: there are more people now than a dying world can support, and the teeming masses of the earth are seemingly more worldly than they have ever been, and are more distracted by pursuits of wealth and diversion than they have apparently ever been, at least since the time immediately preceding the Flood. Sin is rampant in the world, and the ease and luxury humans have gained from the brief respite of Adam's curse have not moved them closer to God. Instead, humans have thrown their sinful and foolish nature into sharp relief by building a thoroughly unsustainable array of complex societies around the world, and have thought themselves to be glorified by them.

God knew what was best for us when He placed Adam's curse on us. The question is whether we have lived in conformity with the relatively simple ways that Christ instructed through His example and through the Spirit of Prophecy, or whether we have tied ourselves to societies that are teetering on the brink of collapse and chaos. Are we, too, reveling in the ease and luxury that the world has so foolishly wasted? The consequences of sin and the foolishness of humanity have laid waste to the global food production systems that God provided, with massive consequences that will be meted out in this current age. Are we as Christians ready to once again gain our food by the sweat of our brow? Are we ready to assist others and adapt our ministries to coming realities? Are we prepared to lead shaken and worried people to Christ? These questions and others are more fully explored in a later chapter. But before we get there, there are several other global systems to look at. Brace yourself: the news is quite bad.

-----

[1] One of the best books I have read on this subject, and one that I highly recommend to anyone wanting to know more is Dale Allen Pfeiffer's *Eating Fossil Fuels: Oil, Food and the Coming Crisis in Agriculture* (Gabrela Island, B.C.: New Society Publishers, 2006).

[2] Pfieffer, pp. 20, 21.

[3] Cooking oil and crude oil have roughly the same number of calories per gallon. See http://caloriecount.about.com/calories-vegetable-oil-canola-i4582?size_grams=218.0.

[4] See www.usda.gov/factbook/chapter2.pdf. Note that this figure is an aggregate amount that includes spoilage, plate waste, and other losses. Actual caloric intake was 2,700 calories.

[5] See www.globalissues.org/article/26/poverty-facts-and-stats.

[6] See www.sciencedaily.com/releases/2009/01/090128074830.htm.

[7] Fuels made from woody plants instead of from such food products as corn. Unfortunately, as this book is written, production of cellulosic fuels is close to zero despite billions of dollars invested in the sector.

[8] Pfieffer, p. 7.

[9] The expansion of population in relation to available food is a well-studied phenomenon. It works with rodents, fish, birds, bacteria and yeasts, and mammals—including humans. After the population burst, when the food is all used up, there is a population crash. The phenomenon is called "overshoot."

[10] See www.theepochtimes.com/n2/china-news/fertilizer-overuse-damages-agriculture-and-environment-in-china-62671.html.

[11] Gretchen C. Daily, "Restoring Value to the World's Degraded Lands," *Science,* July 21, 1995.

[12] *Ibid.*

[13] R. Houghton, "The Worldwide Extent of Land Use Change," *Bioscience* 44, no. 5: 305-313. Available online at www.jstor.org/stable/1312380.

[14] Food and Agriculture Organization of the United Nations (FAO), "Agro-Ecological Land Resources Assessment for Agricultural Development Planning." Retrieved June 29, 2011,

from FAO Corporate Document Repository: www.fao.org/docrep/009/t0741e/t0741e00.htm.

[15] David Pimentel and Mario Giampietro, "Food, Land, Population and the U.S. Economy" (Washington, D.C.: Carrying Capacity Network, 1994).

[16] www.naweb.iaea.org/nafa/news/topsoil-china.html.

[17] Pimentel and Giampietro.

[18] www.ers.usda.gov/publications/aib775/aib775o.pdf.

[19] www.grida.no/publications/et/ep4/page/2633.aspx.

[20] www.arlingtoninstitute.org/wbp/global-water-crisis/606.

[21] www.ela44.instablogs.com/entry/water-crisis-in-pujab-a-global-issue/.

[22] This is an average. Different types of grain crops require different amounts of water. Rice, which is grown in standing water, obviously requires the most, and corn requires the least. All grains, however, require a great deal of water to be productive.

[23] http://green.blogs.nytimes.com/2011/05/04/aquifers-depletion-poses-sweeping-threat/.

[24] See www.Birdflubook.com/resources/epstein747.pdf; www.iiasac.ac.at/Admin/PUB/Documents/XO-02-001.pdf.

[25] See http://people.oregonstate.edu/~muirp/pstlosch.htm.

[26] www.foreignpolicy.com/articles/2010/04/20/peak_phosphorus.

[27] Morocco, China, South Africa, Jordan, and the United States.

[28] www.foreignpolicy.com/articles/2010/04/20/peak_phosphorus.

[29] See www.indexmundi.com/commodities/?commodity=rock-phosphate.

[30] Ellen G. White, *Spiritual Gifts* (Battle Creek, Mich.: Steam Press of the SDA Publishing Association, 1864), vol.4a, p. 155.

[31] *The SDA Bible Commentary,* Ellen G. White Comments, vol. 1, p. 1085.

[32] See Ellen G. White, *Country Living* (Washington, D.C.: Review and Herald Pub. Assn., 1946).

Chapter 5:

The Decay of Our Global Climate System

"Watch ye therefore: for ye know not when the master of the house cometh, at even, or at midnight, or at the cockcrowing, or in the morning: lest coming suddenly he find you sleeping. And what I say unto you I say unto all, Watch." Mark 13:35-37.

When I and my family lived in Mongolia from 1994 to 1999, in the capital city Ulaan Baatar[1] (the only real city in the country at that time), climate change was just beginning to enter the public consciousness in a significant way. Certainly it was a far thing from the minds of average Mongolians, who were dealing with a collapsed economy and with a host of social ills. It was even further from the minds of the hundreds of thousands of nomadic herders who still existed in the country at that time (more than "existed"—they formed the backbone of rural life) and who lived simple lives and followed traditions and grazing patterns that were thousands of years old. It is all the more ironic then that it was these Mongolian herders who were and have continued to be among the hardest hit in the world by climate change, and is especially ironic since these sincere, friendly people live in a way that generates almost no carbon footprint.

Mongolia has experienced climate change in two distinct, firsthand ways: desertification and shifting precipitation patterns. In the first of these, the Gobi desert (located in the south of the country) has expanded to the north at up to 10 miles per year as lands have dried up and deteriorated into desert because of increased temperatures, decreased precipitation, and overgrazing.[2] Families that formerly ringed the northern edge of the desert were forced to change their seasonal migration patterns substantially and move to new winter and summer locations, which placed them in direct competition and conflict with other families that had for their whole lives occupied those lands and used them to graze their herds. The second and far more dire way in which climate change impacted Mongolia was through the advent of winter snow.

I should explain that Mongolia is a very cold country—it holds the record for having the coldest capital city on earth.[3] In Mongolia, winter temperatures of −40°C and −45°C are common enough, but the cold comes with very little snow—usually just an inch or two at most during the course of the winter.[4] This is actually a good thing, because the grass on the plains has been cropped short by the fall, and the animals must range widely to get sufficient food in the winter. By the spring, the vast plains are so grazed down that they are as smooth as a pool table, and the sheep, goats, yaks, cows, horses, and camels that Mongolians herd and live off of are pathetically thin and eager to again see green grass. It is a hard life for both humans and animals, but nomadic herders in Mongolia have survived in these unforgiving conditions for untold generations.

In such a situation, when snow does come (even as little as a few inches), it is a disaster. Animals are forced to remove the snow with their hooves to get to the grass, but there is so little grass that they expend more energy in removing snow than they get back in grass. Unless wind comes and blows the snow away (it will not thaw until late April), the animals get thinner and thinner, then get sick, and then die. And herders who lose their animals lose everything: herders and their families burn freeze-dried dung to stay warm, almost all of their diet consists of meat and milk products,[5] and the walls of their round tents (spelled "*ger*," but rhymes with "bear"—"ghair" would have been a better spelling) are made of wool from their animals. Even the ropes that hold the felt on the ger walls are made from braided horsetail hair. Their animals are their food, their transportation, their clothing, their heat, and their savings.

Starting in the mid-1990s, while I lived in Mongolia, winter snows began getting heavier and coming every year. This was without precedent in the Mongolian collective memory and had a staggering impact. Untold numbers of strong, proud, self-sufficient rural Mongolians lost everything (some even had to burn their scant furniture to try to keep warm for a day or two longer—an act of true desperation) and migrated to the city, where there were no jobs. Some of the desperate displaced rural women turned to prostitution to keep their families from starving, and some of the desperate displaced rural men turned to crime for the same reason. In rural areas millions of frozen livestock carcasses littered the steppes (they were nothing more than skin stretched over bones). For the Mongolians climate change was a very real thing that all but destroyed their rural economy, changed the fabric of their society, created refugees, and created massive suffering and moral decay. Mongolia provides an example of how extensive the impact of climate change can be across a society—even a relatively noncomplex society—from relatively small changes in our climate. Mongolia should serve as a sobering example to the rest of the world.

As real as climate change is in Mongolia and many other parts of the world, it is still difficult to objectively discuss climate change at this point in time because the subject has become so politicized—at least this is the case in the United States. While it is politicized elsewhere (the United Kingdom comes to mind as an example), it seems that people in the United States are perhaps the most polarized on the topic, and wherever there is polarization there is almost always a great deal of inflamed rhetoric and inaccurate information to be found. Fortunately, however, it is not the purpose of this chapter to argue for the existence of climate change through science. Instead, it is the purpose of this chapter to try to find common ground for Adventists on the subject by looking through the lens of the Bible and inspired writings, seeing if there is any basis to expect climate change as one of the effects of the decay of global systems or as a part of end-time events. We will also look at the general effects that science predicts will take place as a result of climate change, and then we will compare these effects to what is generally predicted in Scripture and Spirit of Prophecy to take place in the final days.

There are a few instances in which climate change is implicitly or explicitly mentioned in both Scripture and inspired writings, and, wherever mentioned, it is stated to be caused by sin. Recall that in the second chapter of this book we reviewed a description of the first instance of climate change, which occurred in almost immediate response to sin. In *Patriarchs and Prophets* we read: "The atmosphere, once so mild and uniform in temperature, was now subject to marked changes, and the Lord mercifully provided them with a garment of skins as a protection from the extremes of heat and cold."[6] The context here is that Adam and Eve had just sinned, and the *first* noticeable response to sin in the physical world was a change in the climate. In other words, there was a direct and immediate link between the climate of the earth and sin in the earth. This is an important concept, so let's dwell on it for a moment: The first change in the earth that Adam and Eve note in response to sin (and separation of creation from its Creator) is not the death of a flower or an animal, it is a change in the climate—the impact of sin in something as vast as the earth's climate system was immediate, which indicates how vile and powerful sin is.

It stands to reason that if there was a direct and immediate link between sin and a change in the climate of the earth 6,000 years ago, then there is a very strong chance that there remains a direct link between sin and ongoing changes in the climate of the earth in our age, especially since the effects of sin upon man and upon the earth are progressive and accumulative in nature. As one of my favorite writers put it: "There is no place upon earth where the track of the serpent is not seen and his venomous sting felt. The whole earth is defiled. . . . The curse is increasing as transgression increases."[7] Thus, we know that the effects of sin upon the physical world increase through the ages, and we know that the first impact of sin was upon our climate. But let's not connect the dots quite yet—there is more to add to our study.

The second instance of climate change can be found through implicit reference in Genesis 7:10-12: "And it came to pass after seven days, that the waters of the flood were upon the earth. In the six hundredth year of Noah's life, in the second month, the seventeenth day of the month, the same day were all the fountains of the great deep broken up, and the windows of heaven were opened. And the rain was upon the earth forty days and forty nights." Consider for a moment the fact that before the Flood there was no rain—God created the world with a climate system that did not produce rain. Also consider what change was necessary to God's original climate system to cause untold billions of tons of water to be wrung out from the atmosphere as rain. We know that God caused the Flood to happen (Gen. 6:6-8) and that it was neither a random event nor an early result of the decay of global systems. We don't know what agent God used to change the nature of the earth's atmosphere (God could have changed the climate to precipitate rain, or could have used another mechanism to precipitate rain and therefore change the climate), but we do know that truly awesome amounts of water were precipitated onto the earth,[8] thus forever changing the earth's atmospheric system[9] and climate and also forever changing the nature of the earth's crust ("the fountains of the great deep [were] broken up").

God dramatically changed the climatic system of the earth when He caused the Flood to occur. And He did it in direct response to sin: "The earth also was corrupt before God, and the earth was filled with violence. And God looked upon the earth, and, behold, it was corrupt; for all flesh had corrupted his way upon the earth. And God said unto Noah, The end of all flesh is come before me; for the earth is filled with violence through them; and, behold, I will destroy them with the earth" (Gen. 6:11-13). The linkage between sin, Satan, the climate of the earth before the Flood, and the actions of God are all made crystal clear in *Patriarchs and Prophets:* "In the condition of the world that existed before the Flood they saw illustrated the results of the administration which Lucifer had endeavored to establish in heaven, in rejecting the authority of Christ and casting aside the law of God."[10] While the decay, deformity and death of man under the rule of Satan was rapid and willful, the administration of Satan has had no less a certain effect upon the rest of earth. Simply put, decay and death are the natural result of rejecting the law of God and separating creation from its Creator. And of course, this law of decay and death extends to our climate system as surely as it does all the other global systems that God created.

So here, based on the Bible and inspired writings, is what we know about climate change:

- Separation from our Creator causes decay, deformity, and death. This is the result of Satan's administration.
- The *first* effect of sin upon the physical world that was noticeable by Adam and Eve was a change in the climate.
- God changed the climate of the earth when He caused the Flood.
- The effects of sin upon the earth increase "as transgression increases."

With sin and the curses of God weighing on it, the post-sin, post-Flood climate system we now have in place is a poor shadow of God's original design. And this poor climate system is continuing to be affected both by the accumulated sin of the ages, as well as by the exploded population and sin-induced excesses of our current industrial age. Seen in this context, it is not surprising that our climate would be changing as part of the last-day events. In fact, the truth is just the opposite: Given the established link between sin and climate, it would be surprising if the earth's climate was *not* changing in these, the last sin-filled days of the earth.

Climate change may be a politically charged subject in these, the final days of earth's history, but it is hard for people on either side of that debate to argue against the preponderance of information in the Bible and inspired writings that strongly indicate that climate change will be one of the consequences of decay in the global systems of the earth. This is particularly true since there is a strong correlation between the impacts that science generally expects because of climate change—such as "extreme weather"—and the events that are expected in the last days as outlined in the Bible and inspired writings. On page 58 we look briefly at this correlation, but since this book tries to point to the imminent fulfillment of the state of the world as described in Matthew 24:6-8 and since the physical impact of climate change is only a small part of that argument, this chapter will spend more time looking at the impact of climate change on the societies of humankind, and how these impacts will bring us closer to the state of the earth as decribed in Matthew 24 and the book *Last Day Events*.

In looking at climate change correlations of last-day events in the Bible and inspired writings, we need to be careful that we consider cause. It is true that the systems of the earth are decayed, deformed, and dying as a result of sin. However, it is also true that, after a certain point, the war between Christ and Satan will be resolved and the end result of Satan's administration will be conclusively known and forever settled, leaving Satan with no effective argument against God. At this point in earth's history, God takes a much more active hand in end-time events, and the pace of events increases dramatically. Because of this, this book will not use weather/climate events

Source	Impact	Correlation With Expectations of Science
Matthew 24:7 and *The Great Controversy*, chapter 36	**famines**	Decay and disruptions in food production, water availability, oil availability, and crop disasters because of disease and pestilence will reduce food production significantly, while war will disrupt delivery. See chapters 4, 5, 6, 7, and 8 in this book.
Matthew 24:7 and *The Great Controversy*, chapter 36	**pestilences**	Climate change is expected to bring significant threats in the form of pests and disease, including pests and diseases that are new to regions that are experiencing climate creep. See chapter 4 of this book.
The Great Controversy, chapter 36	**calamities by sea and land**	Higher levels of energy (heat) in the oceans and atmosphere are expected to lead to higher levels of storms ("extreme weather") that will result in increased losses of property and life on land and at sea.
The Great Controversy, chapter 36	**fierce tornadoes, terrific hailstorms**	Higher levels of energy (heat) in the atmosphere are expected to lead to more intense storms, including more intense tornadoes and hailstorms.
The Great Controversy, chapter 36	**tempests, floods, cyclones**	See above.

from Revelation to buttress its argument but will instead use quotes from *The Great Controversy* and *Last Day Events*, since these are more expansive and are more readily identifiable on a time line of final events.

The table on page 58 includes a few select quotes from the Bible and *The Great Controversy*, and is meant to illustrate how impacts from climate change resulting from the sin-induced decay in our global climate system play a role in end-time events.

There is an abundance of readily available data on the expected impacts of climate change, and, since the earth and its systems are being more intently and intensively studied by more scientists than have ever studied anything, the information is frequently updated. With this in mind, it is not the purpose of this book to repeat currently available climate science or impact analysis. Instead, this chapter will deal more generally with the impact of climate change on the societies of humankind and how these impacts promise to bring the world much, much closer to the state predicted in Matthew 24 and in inspired writings. Let's start our inquiry by looking at why climate change is expected to have a broad impact on human societies.

Over the past several years there have been heard not infrequent comments from people who questioned whether the impacts of climate change would really be so bad. "Would a couple more degrees of heat really be such a big deal?" they would ask. Or: "A melting arctic just means more room for more crops, or an ice-free Northern Passage, right?" Such views toward climate change are quite simplistic and completely miss the major risk of climate change, which is this: climate change puts into doubt most or all of the assumptions upon which our complex industrial societies around the world are built. Let that sink in. Under a climate-change scenario, our societal assumptions can no longer be taken for granted. In other words, our cities, farms, roads, bridges, dams, levees, electrical grids, and water distribution systems, etc., were all located and constructed where they are and how they are under the assumption that the conditions that prevailed when they were established would continue without significant change. It is hard to overstate what a big deal this is. With climate change, storms, droughts, heat waves,[11] pestilences, and disease that did not previously affect our societies are suddenly making a profound impact. Put another way, climate change destabilizes societies on a global basis by significantly increasing costs and increasing disruptions and other risks while simultaneously and dramatically decreasing societal productivity and food production. Our societies, already under tremendous pressure from our burgeoning global population and from poor management of our finances and natural resources, cannot afford the increased risks (both known and unknown) and disruptions that climate change brings. We don't have enough resources to rebuild our infrastructures on a global basis, and only the rich nations will be able to afford to retain a shadow of their current prosperity and ease, leaving billions to intense suffering and unknown millions to death.

That was a pretty intense rush to the heart of the reason for concern on climate change. Let us go back a bit and cover the ground more slowly and methodically this time.

Our human societies, dating in some instances from not long after the Flood, have either been established to take advantage of the resources around them (situated in ideal agricultural settings, or situated at the edge of rivers or oceans, for instance), or have been situated in places that were near resources but were out of the way of harm (situated above flood levels, or beyond the risk of an avalanche, etc. In the case of North America, significant populations and agricultural activities came quite late in the context of the history of earth, and yet in each place that societies were established, they were established to take advantage of the resources surrounding them. And in each case, the establishment of a settlement was underpinned by a multitude of assumptions. Settlers found a good spot for their houses and then assumed, for instance, that their settlement would not be swept away by frequent massive floods. They also assumed, based on the plant life they observed, that there would be sufficient rain for their crops, and they assumed they would get enough sun to be able to raise crops or, in the case of ocean settlements, that they would be able to harvest food from the ocean and use boats or ships to trade for other needs. Not all assumptions were correct, and not all settlements flourished, but where mistakes were made, settlements, or at least settlers, were moved to a more hospitable area.

As surviving settlements grew, they continued to make assumptions, including assumptions they did not know they were making. They extended their agricultural activities, assuming that fields would continue to produce and that the climate would remain conducive to growing the crops they were planting. Settlements, and later towns, cities, and societies as a whole continued to expand and increase investments in infrastructure (roads, water systems, waste systems, power systems, etc.) while increasing significantly in terms of complexity and interdependence.

This last point is critical to understanding how fragile and susceptible to climate disruptions our modern societies are. Interdependence means that industrial operations (paper mills, bakeries, cement factories) are reliant upon inputs (such as trees, wheat, or rocks) they do not control in order to achieve their outputs (such as paper, bread, and cement). Factories are reliant upon power, for instance, and often upon water. They are certainly reliant upon fuel, both to get their inputs and to ship their outputs. In complex and interdependent societies such as now fill the world, any breakdown in any part of an industrial society has severe consequences for the rest of the society, but especially the "downstream" parts of the society. For instance, let us say that a water main breaks in a medium-size city, flooding a main road, causing a sinkhole under some railroad tracks, and leaving a fair portion of the city without water for a couple days.

Any city dweller who has lived through some variant of this scenario can predict what will happen: without sanitary facilities, affected schools and offices will close or adapt (perhaps by bringing in their own water at great expense), factories who are reliant on water will be forced to shut down or adapt (usually at great expense), and transportation linked to the railroad will be shut down for a period, wreaking havoc with downstream factories that are waiting for their "just in time" deliveries of goods. So shut-down factories in one city can affect other factories in other cities and other countries, passing along some measure of instability as a result of even minor disasters.

One broken water main can disrupt events significantly on a local basis but not have a great overall impact. However, if the scale of disruption was equal to two broken water mains in half the cities of the world, the effect would be felt globally, and costs would rise while productivity fell. Now imagine the effects of a dozen broken water mains in three quarters of the cities of the world, and you will begin to get a feel for the scale of impact that can be compared to the massive disruptions in the natural world that are associated with climate change. Imagine the impact of a dozen simultaneous major floods around the world, or a series of large hurricanes on America's East Coast combined with several significant droughts in America's Southwest and West, combined with a couple killing-scale earthquakes on the West Coast for good measure. Such a scenario is an end-time reality, and if we take a good look at the complex societies that fill the globe, and if we begin to evaluate more closely the weaknesses and fragility that are a part of those societies, we see that our complex societies do not tolerate disruptions in critical services, nor are they able to readily adjust to disruptions in a flow of supplies. Decreases or interruptions in the availability of power, water, sewerage, and waste-hauling services, fuel, food, and goods and supplies bring complex societies rapidly to their knees.

As this book is written, global societies are more complex and interdependent than they have ever been at any time in recorded history.

Volcanoes in Iceland, for instance, can impact the functioning and productivity of all of Western Europe. February temperatures in Florida and July rains in the Midwest United States can affect the price of food on a global basis. Our global economies are now deeply integrated, and this seeming strength becomes a key weakness as we find that, in an age of increasing disasters, an instability anywhere in the world is an instability everywhere in the world. As this book was being written, Europeans were learning hard lessons about a weakness anywhere in the European Union being a weakness everywhere in the European Union. It seems safe to predict that more such lessons along these lines are in store for humankind.

Science predicts that global uncertainty and risk will increase as a result of climate change.[12] For instance, we will get more floods, and they will be more severe,[13] with each flood having a local, regional, national, and perhaps international consequence in regard to costs and potentially the disruptions in the flow of goods. At the same time, we are expected to have more and more powerful hurricanes throughout the world,[14] each of which delays shipping and some of which cause huge damage to societies and their infrastructure. Heat waves and droughts, which are already on the rise and which are expected by climate scientists to increase significantly in the immediate future,[15] also have significant impacts on food production, which of course has dramatic ripple effects across societies, causing discontent and, in severe cases, social unrest and upheavals. Pestilences and diseases that attack plants, animals, and humans are expected by scientists to increase dramatically under climate-change scenarios.[16] One impact of such pestilences and diseases is that they will decrease food production wherever they strike, and with decreased availability of food will come increased cost. In sum, disasters will dramatically increase societal costs, dramatically decrease societal stability, and dramatically impact local, regional, and national budgets.

Note that all of the effects that science expects from climate change work together to destabilize societies and economies, and that is a grave threat to the billions who live on the earth, especially the very, very poor who are already barely surviving, as is currently the case with 3.5 billion people who live on $2.50 or less per day. Add to this the effects we have already studied regarding decreased food availability and increased food prices, and we can see that it is upon the poor that the events of this, the final age of the earth, will fall most heavily.

We cannot stop what is coming—it is the result of 6,000 years of accumulated sin and is the culmination of the war between Christ and Satan and the fulfillment of prophecy, and will not be stopped or even slowed. Because of this, the peril of these billions of souls raises crucial questions: Will the billions who are either already in crisis or who are steadily progressing toward crisis die without hearing the gospel? What should be the course of action for our church in regard to the unevangelized billions? What should the course of action be for ADRA and Adventist Community Services given what we know is coming? How will we use Adventist World Radio in this historic time? How will we use individuals willing to be missionaries and witnesses? Moving from the corporate to the personal, what should the course of action be for you and me in light of these grave needs and the greater challenges that are coming? We will discuss these issues in some detail in the final chapters of this book.

In the year 2011—the writing of this book was completed in December of 2011—a record was set: natural disasters the world over were running at a record pace,[17] with natural disasters in the United States costing more than $1 billion happening every month through September. Global disaster losses in 2011 were running in the hundreds of billions—a pace of loss that, if maintained, would weigh heavily on even the most resilient of the world's economies. If the descriptions of end-time disasters in the Bible and in inspired writings can be applied to any year thus far, it would be 2011. And if they can be applied to any age, it would be the one we are in. Surely we are seeing the birth pangs. Unfortunately, another prophecy is also true—much of God's church in this age is Laodicean and has not awakened, even in response to the signs that are all around us.

In this chapter we covered the decay of God's global systems through the lens of climate change, because that lens is so overarching and so clear. But there are other global systems created by God that we have yet to cover, and the next one is the decay in our oceanic system. The news, unfortunately, is all bad.

---

[1] Now more commonly spelled "Ulan Batar," but I am attached to the spelling that predominated when I lived there, and I do feel that the two words that make up the name of the city ("ulaan," meaning "red," and "bataar," meaning something between "gentleman" and "hero") should be preserved in the spelling of the city name.

[2] See www.swiss-cooperation.admin.ch/mongolia//resources/resource_en_158826.pdf.

[3] See http://geography.about.com/od/physicalgeography/a/coldcapital.htm.

[4] See this brief but excellent explanation of Mongolian weather and temperature statistics, and the dynamics that influence precipitation patterns: www.koreanhistoryproject.org/Jta/Mo/MoWX0.htm.

[5] The Mongolians eat a wide variety of meats and have many ways of preparing or preserving the milk of various animals. While traveling in the Gobi desert with my family, I

had a very memorable bowl of Cheerios (mailed halfway around the globe to our kids by their grandparents) with camel's milk over it. I cannot recommend it.

[6] E. G. White, *Patriarchs and Prophets*, p. 61.

[7] *The SDA Bible Commentary*, Ellen G. White Comments, vol. 1, p. 1086.

[8] And hereby "the waters above" came to hold much less water after 40 days and 40 nights of rain.

[9] We know that an atmosphere with less water is less capable of storing energy and is a less effective buffer to thermal change and a less efficient distributor of thermal energy. An earth with the capabilities of its original atmosphere intact would very likely be a place where the poles of the earth were much warmer than they currently are and much more hospitable to plant and animal life.

[10] E. G. White, *Patriarchs and Prophets*, pp. 78, 79.

[11] In the United States, Texas in the summer of 2011 may be the best recent example of what a climate-change-induced heat wave could look like and how broad its impact may be.

[12] See www.justmeans.com/New-Report-Uncertainty-Increases-Risk-in-a-Changing-Climate/23933.html and www.guardian.co.uk/environment/2011/jul/06/climate-change-war-chris-huhne.

[13] www.physorg.com/news/2011-02-driven-climate.html.

[14] www.usgcrp.gov/usgcrp/links/hurricanes.htm.

[15] See www2.ucar.edu/news/2904/climate-change-drought-may-threaten-much-globe-within-decades.

[16] http://crisisboom.com/2011/07/01/farm-animal-disease-to-increase-with-climate-change-scientists-say and www.standardsfacility.org/files/ClimateChanges/STDF_Coord_292_Session1_MatthewAryes_Sep09.pdf.

[17] www.msnbc.msn.com/id/43727793/ns/world_news-world_environment/t/already-costliest-year-natural-disasters/.

Chapter 6:

The Decay of Our Oceanic Systems

"In the last scenes of this earth's history, war will rage. There will be pestilence, plague, and famine. The waters of the deep will overflow their boundaries. Property and life will be destroyed by fire and flood."
—*Ellen G. White,* Maranatha, *p. 174.*

My wife and I were married in Sterling, Massachusetts, in August of 1986. We honeymooned in Maine, wandering up and down the coast of that state without a plan or a schedule, staying at random bed-and-breakfast places that caught our attention. We loved it.[1] Early on in our honeymoon we were staying in Kennebunkport, with an ocean view from our room. I noted that only a block from where we were staying there were local charter boats offering deep-sea fishing, and I suggested to my bride that we try going out on one of the boats since the weather was warm enough that a day on the sea would be welcome. And so we boarded a small boat—maybe a little more than 25 feet—along with perhaps six other customers and the captain. As we cruised out of the harbor, the captain, who seemed to be about 50 or so,[2] started telling us stories of how, as a boy, he would take a little rowboat out to a certain buoy in the harbor and would tie his boat to the buoy, drop his line, and quickly catch as many fish as his family could use.

The captain told the story with more than a trace of sadness in his voice, and said that things in the Gulf of Maine had changed tremendously since he was a boy, and that we would be looking for the same fish he had once caught in his little rowboat, but that we would have to go out about 10 miles and would be quite lucky to catch anything at all. The day consisted of going from spot to spot and fishing for a little while without getting a bite. After the first couple of spots my wife and I didn't bother to fish but instead enjoyed the sun and listened to the captain, who continued to talk, with evident pain, about the problems and decline in the gulf and the disappearance of once-plentiful fish. The captain would stare out at the ocean, as if seeing something that was no longer there, and talk about his father and his uncles, who had all been fishermen and who'd all had their own boats. He spoke of the great quantity of fish that they'd caught, and how they had made a life for themselves from the abundance of the sea, along with so many other men who worked on the Gulf of Maine. It was obvious that the captain not only felt pride in his roots and family history, but also loved the gulf, was mourning for it, and was embarrassed for what had become of it in only two generations. His embarrassment was likely heightened by the fact that he could not find a single fish for his customers to catch in what had been, only a couple decades before, an extremely productive area.

Once during a lull in his stories the captain pointed at a large ship several miles away and said, "That is part of the problem right there. Not the whole problem, but for sure they are part of it." Naturally the passengers took the bait and asked the captain what kind of boat it was. The captain spit in the water before replying. "That is a bottom trawler. They scrape every last living thing off the bottom, a hundred yards wide with each pass, and bring it all up. They leave a desert behind them where nothing can live." He shook his head. "It is only a part of the problem, but I still can't believe they let them do it."

That melancholy captain and his stories have stuck with me for more than 25 years, and during those years I have found that he is not alone. I have gained the impression that all over the world and in almost every port in every sea there are legions of older fishermen who speak wistfully of the days when the oceans were healthy and the catch was good, and for good reason—the quality of our oceans has been more or less in steep decline for the past 50 years.[3] During this time there have been profound changes in ocean chemistry, water temperature, the viability of shallow-water "nurseries," and the rate at which sea life is harvested.

The steep decline in the health and viability of our oceans is reason for serious concern. For one thing, our oceans are the primary source of protein for about 1 billion people—people who have few other protein options. Of even greater concern is the fact that our oceans so strongly affect the other global systems that keep our world functioning. If our oceans are profoundly changing, then we can expect all of the systems upon which we depend for life, water, and food to decay even more rapidly. In other words, the steep decline of ocean health and function is a strong sign that we are progressing rapidly toward an earth as described in Matthew 24:6-8.

Our oceans are big. Really big. When I was living in Mongolia and traveling occasionally between Asia and the United States, I flew across the Pacific Ocean on a regular basis. It never ceased to amaze me that I could board a plane in Los Angeles, be over the Pacific Ocean within 15 seconds after takeoff, fly at more than 500 miles per hour for 12 hours, and *still* be over that same ocean. For sin-filled human beings to be able to so pollute and alter and pillage our massive oceans that they are in a state of steep decline—and do most of it in the space of only two or three generations—is a failure of stewardship that is unparalleled in the history of humanity. And yet it is also an act that is consistent with the nature of selfish and sin-filled human beings.

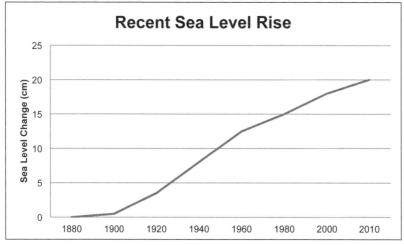

Sea levels are steadily rising.
Source: Permanent Service for Mean Sea National Oceanography Centre, Liverpool, England

How bad is the state of our oceans? People who study the oceans have come up with a term to name what they say is the coming state of the seas—the "Myxocene," or the "Age of Slime." They name it this because, at the rate we are polluting, toxifying, and overfishing the oceans, there will soon be nothing left in them but slime—pretty much the only thing selfish humans cannot eat or otherwise profit from.[4] What is striking is that for thousands of years the abundance of the oceans was thought to be inexhaustible. Indeed, as late as 1967, when the global sea harvest (everything taken from the sea, not just fish) was about 50 million tons, the U.S. Interior Department predicted that the harvest could be increased at least fortyfold, to 2 billion tons per year, without endangering aquatic stocks.[5] As it happens, global sea harvest peaked in the late 1980s at about 85 million tons[6] and has been declining ever since, despite better and better technology being used to find and catch fish and despite ultra long-range boats being used to fish the most remote waters in the world.[7] The U.S. Interior Department was way, way off in their projections of the amount of resource that was actually available—a common mistake that we seem to repeat with oil, minerals, crops, water, and all manner of resources.

The reasons for the global decline in fisheries and in the overall health and function of the oceans are complicated, partly because the chemical and thermal interactions that the oceans have with the rest of the planet are so active and complex, but mostly because the oceans are and have been under assault throughout the industrialized age,[8] and are affected by pretty much everything that humanity does on the globe, even when it takes place thousands of miles from the ocean. Once again we return to the fact that all of God's created systems are very interconnected and interdependent—what happens with one system has repercussions in all other systems, and as one system decays it contributes to and hastens the decay of all other systems.

Because oceans are connected to every other system, it is hard to know where to start in describing the sources of their decay. It is also hard to know where to stop. Just the same, let's take a look at some of the major incidents of decay in the oceans of the world and sketch out the impacts of those incidents of decay. Since oceans have an array of functions in God's created systems, let's look at the decay of the oceans through the lens of their functions. We'll just skim the surface of the oceans, so to speak, but even doing that is illuminating and sobering.

Gas Exchange and Host to Vast Numbers of Living Creatures: The oceans of the world interface with the atmosphere and continuously exchange vast amounts of energy and chemicals in the form of heat and gases/aerosols.[9] This process is a significant part of what drives planetary weather and variations in this process explain, in part, regional disparities in climate and weather. One of the gases that oceans have absorbed in massive quantities in the industrial age is carbon dioxide—CO_2 (which is readily absorbed by cool water and can be given off by that same water when it warms). Because oceans initially absorbed our excess CO_2, global warming has been slower than early models predicted. However, while models have improved, what was slow to be recognized is that the absorption of CO_2 in the ocean creates carbonic acid, thus acidifying the ocean. Recently oceanic acidification has accelerated at an abrupt and alarming pace, causing observed degradation in some shellfish (whose shells are now slowly dissolving) and causing scientists to assert that the oceans may be passing a critical CO_2 threshold.[10] This is a big deal: the ability of the oceans to buffer CO_2-induced acidity has been overwhelmed, and a loss of oceanic life is already occurring and will increase in speed and scope. Corals (the "nurseries" of many fish species) are already being impacted and are expected to be extinct by mid-century.[11] It is hard to overstate the impact that a loss of coral will have on the rest of the life systems in the ocean, but it is certain that fish harvests will decline steeply as the "nurseries" of the ocean die out. The seas will likely continue their gas exchange functions in our age, but in doing so they are becoming an increasingly hostile environment for the enormous range of life-forms that live in them, and a far cry from what God created.

Distributor and Mediator of Thermal Energy: The oceans are a global distributor and mediator of thermal energy—heat—where surface water is warmed in the tropics and is drawn toward the poles to replace water that has cooled, condensed, and sunk. The cool water that has sunk is in turn pulled upward in tropical regions, replacing water that was drawn north and helping to moderate temperatures in the tropical belt. The entire system is essentially a conveyor belt running on heat. There is a great deal of study and speculation on the topic of whether or not climate change will slow or stop these critically important currents. There is no ready way to calculate the monetary impact of a significant change in thermal distribution around the globe, but the impact will be profound, as cool places will get cooler and warm places will get much warmer, and agriculture and industry will be seriously disrupted as a result. Some climate change models predict the possible collapse of the ocean's thermal energy distribution system while others call for a possible slowing of the system. The science on this subject is not at all settled, but one thing is certain: Any change in the ocean thermal distribution system will be very negative—even apocalyptic—for the rest of the earth and will affect everyone, no matter how far from the ocean they live.

There is, however, another side to the oceans as thermal distributor and mediator for the world, and it is this: The heat that is absorbed by the oceans is energy. Some of that energy is transported to warm cooler regions of the earth, some heats the seas themselves, and some is released into the atmosphere through storms, such as hurricanes. When the seas absorb more heat, such as they do under climate change conditions, then the storms that are generated by the oceans are more frequent and more severe. This is just simple math, but the implications of the resulting "extreme weather" storms for the world are huge and are consistent with the predictions found Matthew 24:6-8. More and larger storms mean more death and more destruction and more spread of disease. They also mean more flooding and more erosion of land, which means reduced crop production and increased costs related to maintaining our infrastructure—our roads and bridges and pipelines, etc. As well, more and larger storms mean the spread of more pestilence, which again means reduced crop production. The cumulative cost to society of more and larger storms is sufficiently massive to be destabilizing, especially when taken in the context of all the other impacts from global system decay.

Another interesting side to oceans as thermal distributor is that water expands ever so slightly as the oceans absorb more and more thermal energy (or, in simpler terms, as the water gets warmer it expands). The amount of expansion is extremely small, but when combined with the additional water pouring into the oceans from melting glaciers and ice caps, the result is increasing sea levels that are even now forcing some island nations, such as the Maldives (the lowest-lying nation on earth), to consider moving all their citizens to land purchased from another country.[12] The whole situation brings to mind the *Maranatha* quote that headed this chapter and that, until the science behind climate change began to predict rising sea levels, was puzzling: "In the last scenes of this earth's history war will rage. There will be pestilence, plague and famine. *The waters of the deep will overflow their boundaries.* Property and life will be destroyed by fire and flood"[13] (emphasis added). With ocean warming and glacial melting already underway, the quote is puzzling no longer.

Chemical Recycler for the World: When rain falls on the earth and minerals and organic compounds are leached from the land and from plants, those minerals and organic compounds flow into streams, which flow into rivers, which flow into the ocean. A healthy ocean uses the minerals, nutrients, and organic compounds flowing into it for food and to support biological and biochemical processes and ends up, over the ages, recycling the minerals and unused organic compounds.[14] Nutrients and minerals commonly end up in water. For instance, when rain falls on dead corn stalks or dead grass or other organic material, it leaches out alkali metals such as potassium and calcium and magnesium, some of which is absorbed by the land and some of which stays in the water that runs off into creeks and rivers and eventually the ocean. And when rain falls on farmland, it also absorbs and carries away from excess fertilizers nutrients and minerals such as nitrogen and phosphorus. Some flow of minerals and nutrients from the land to the sea is necessary and healthy. But when synthetic fertilizers are applied in vast amounts and are flushed out to the sea each year (along with other chemicals and sewage residues and street washings, etc.), they have a powerfully destructive effect. The nutrients washed out to sea encourage algae blooms, which in turn grow quickly, then die and sink. When bottom-living bacteria digest the vast blooms of dead algae they use up all the oxygen in the water, resulting in oxygen-starved lower-levels of water called "dead zones." Nothing that needs oxygen can survive in a dead zone. The massive amounts of fertilizer washed down the Mississippi River each year has resulted in a dead zone in the Gulf of Mexico that has roughly doubled in size each of the last several decades and in 2010 was almost the size of New Hampshire. There has been a 300 percent increase in the amount of fertilizers and nutrients washed down the Mississippi River since the 1960s[15] and, of course, a corresponding increase in the size of the dead zone in the Gulf of Mexico. But dead zones are not found just in the Gulf of Mexico. At last count there were more than 500 dead zones around the world (all from fertilizer and industrial run-off) and more than 200 additional potential dead zones beginning to form.[16] This total is up very significantly over just the last decade.[17] It is interesting to ponder that poisonous "red tide" phytoplankton have a complex relationship with dead zones, and that oxygen-depleted water can help spur instances of "red tide."[18] While dead zones are below the surface of the ocean, if the "dead" but still nutrient-rich water were to be cycled to the surface through storms or some source of trauma (such as an earthquake or comet strike), it is possible that it could spur tremendous growth in "red" bacteria, making the ocean appear to be made of blood.

Seas as a Food Source for Millions: According to UNEP, about one billion people get most of their protein from the sea.[19] Given the global pressure on food supplies worldwide, sea protein is irreplaceable. And yet the availability of certain food stocks (types of fish) from the ocean has declined

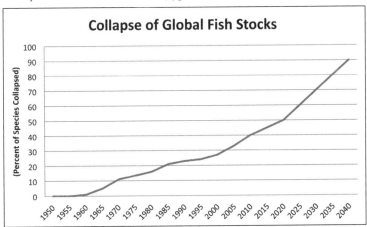

Global fish stocks are collapsing, and our oceans as we know them will soon be destroyed. Data Source: Science/FAO

significantly (see p. 71) over the past several decades while at the same time harvests of fish and other seafood has remained high. How can we keep a high level of seafood harvest while fish population decreases sharply? The answer is fairly straightforward, and it involves three primary reasons. Reason one is that we are employing sophisticated technology in finding, catching, and processing fish. We can now find and capture fish that we could not get to in the past. The second reason is that, globally, we have changed our standards and our tastes as we have depleted traditional fish species and have moved on to species once considered too poor for the table, or have started eating fish that were once too deep or too remote for us to catch.

We have, in essence, been fishing down the food web, moving from larger, top tier species to smaller species that, when taken, disrupt the balance of the oceanic food chain.[20] The third and final reason is that humans have, throughout much of the globe, so aggressively pursued commercial fish stocks that they have caught not only the breeding fish, but in many cases have fished down populations of juvenile fish as well. Fishing out a breeding stock makes it so that only a handful of juvenile fish remain to try, once they reach breeding age, to rebuild the population. Our industrial fishing fleets move from fish stock to fish stock, significantly reducing them or wiping them out as they go, in order to maintain a high level of catch. More effort in the form of better gear and bigger and better boats means more catch—for a short while. In the end, more effort means no more fish. Illustrating this point, the graph below shows the relationship between effort and catch (graph courtesy of Worthwhile Canadian Initiative).[21]

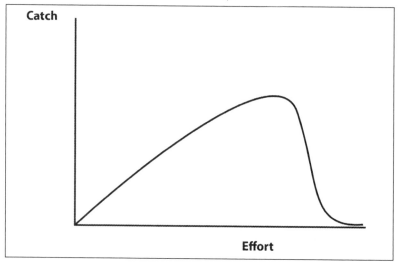

It is not just overfishing that is causing fish stocks to decline. The oceans are becoming a more and more hostile and difficult place for fish to live because pollution, warming waters, increased disease, decreased foodstocks, and acidification are all making it more difficult for fish to survive in the ocean. In this environment, it seems that a species that has been fished down stands very little chance, if any, of ever recovering its former robust population. With all these pressures, scientists now foresee a steep decline in global fisheries, culminating in a total global collapse before 2050,[22] with small species collapsing a bit faster than predator species.[23]

The cost to human societies of the steep decline in our ocean system is gravely high in many areas, but the most noticeable of these is the impact on our global food supply. Consider that the sea has long provided relatively low-cost protein to the developing world and that a large percentage of the millions who rely upon fish as part of their diet cannot afford to pay for higher-priced protein such as results from aquaculture and long-range fishing fleets. Once again we see that it is the poor who will suffer terribly and who will perish in large numbers because of the famines that the Bible tells us will be prevalent in our age.

From a worldly perspective the question naturally arises, "What can we do to save the oceans?"[24] The answer, for Christians, is "nothing," since it is sin and the administration of Satan that is killing the seas and, in fact, the earth, and only the coming change in this administration will effectively address the problem. But that won't stop the world from trying, if only to attempt to avoid the costs of continued ocean system decay. The problem is, human society does not have the technical or financial resources to undertake the effort. We have freely and often carelessly used the resources and services of the sea without truly valuing them, only to discover that the replacement value of the goods and services that the oceans gave us is well beyond our reach.

We don't know how much it would cost to try to restore some small corner of the oceans. The landmark Stern Report[25] was issued in 2006 and was a good first attempt to address the costs of mitigating climate change. Unfortunately, the Stern Report was considered outdated shortly after its release because of both climate change and accumulating scientific data moving more quickly and more pervasively than was anticipated by Stern and his team. Subsequent modeling suggests that the value of lost environmental services (much of which comes from the oceans) will amount to up to 7 percent of global GDP, or an annual cost of about $4.4 trillion.[26] This does not include damage and loss from storms or topsoil erosion or reduced productivity—it is just a valuation of the value of reduced environmental services, and the oceans are key to providing these services.

The bottom line for the oceans is simple: They are doomed because of the accumulated effects of sin in all ages including ours, and the impact of their accelerating decay and death will be massive. There is no hope of reviving them or reversing their course, and their accelerating decay will hasten decay in all other global systems. The continued decay of the seas will directly and significantly affect more than a billion people that live on or near coasts, imperiling their food supply and requiring radical adaptation and perhaps even forced migration. Taken with everything else we have learned about the decay of God's global systems, it is easy to see the decay of the oceans as a powerful and slow-moving but accelerating force that brings us closer every day to the state of the world just before the time of trouble as described by Christ.

We have only touched on the conditions that contribute to the decay of the ocean and the impacts of that decay, but let's move on to the decay in our global freshwater system, where, again, the news is uniformly bad: the extent of decay in God's global freshwater system rivals that of the ocean.

---

[1] We must have really liked it, for 15 years later we made our home in the state, though in the far west of the state, well removed from the ocean and from most tourists.

[2] He seemed old at the time, but now I realize he was not old at all—pretty much just hitting his stride, actually.

[3] Elizabeth Kolbert, "The Scales Fall," *New Yorker,* August 2010.

[4] Daniel Pauly, *Five Easy Pieces* (Washington, D.C.: Island Press, 2010).

[5] Kolbert.

[6] *Ibid.*

[7] For more information on overfishing, read Charles Clover, *The End of the Line* (New York: New Press, 2006).

[8] For a fantastic book that explores what the oceans were like prior to the industrialized age (in other words, just how seriously they have declined from a perspective that is more than generational), see Callum Roberts, *The Unnatural History of the Sea* (Washington, D.C.: Island Press, 2007).

[9] David Herring, "Ocean and Climate," *Nasa Earth Observatory.* Available online at earthobservatory.nasa.gov/Features/OceanClimate/.

[10] S. Leahy, "Climate Change: Oceans Passing Critical CO_2 Threshold." Available online at http://ipsnews.net/news.asp?idnews=44836.

[11] *Ibid.*

[12] www.csmonitor.com/Environment/Bright-Green/2008/1111/faced-with-rising-sea-levels-the-maldives-seek-new-homeland.

[13] Ellen G. White, *Maranatha* (Washington, D.C.: Review and Herald Pub. Assn., 1976), p. 174.

[14] This happens through a myriad of very interesting processes, not the least of which is subduction through tectonic plate movement and subsequent ejection through volcanic action.

[15] www.sciencedaily.com/releases/2011/06/110615091057.htm.

[16] See the World Resources Institute interactive map of dead zones at www.wri.org/project/eutrophication/map.

[17] www.classroomearth.org/node/1794.

[18] www.tulane.edu/~bfleury/envirobio/enviroweb/DeadZone.htm.

[19] www.unep.org/dewa/pdf/Environmental+Consequences_of_Ocean_Acidification.pdf.

[20] www.seaweb.org/resources/briefings/fishdownweb.php.

[21] http://worthwhile.typepad.com/worthwhile_canadian_initi/2011/01/the-generosity-collapse.html.

[22] www.nytimes.com/2006/11/03/science/03fish.html.

[23] www.pnas.org/content/108/20/8317.full.

[24] Again, from a worldly perspective, the oceans are probably the starkest example of "the tragedy of the commons" that this world has ever seen and, therefore, a clear-cut victim of sin and selfishness in humans.

[25] www.hm-treasury.gov.uk/sternreview_index.htm.

[26] Dorothée Herr and Grantly R. Galland, *The Ocean and Climate Change* (Gland, Switz.: International Union for Conservation of Nature and Natural Resources, 2009), p. 53. http://data.iucn.org/dbtw-wpd/edocs/2009-039.pdf. See also Google's public data link on global GDP, which on January 1, 2012, was $63 trillion. www.google.com/publicdata/explore?ds=d5bncppjof8f9_&met_y=ny_gdp_mktp_cd&tdim=true&dl=en&hl=en&q=world+gdp.

Chapter 7:

The Decay of Our
Freshwater Systems

*"O Lord, to thee will I cry: for the fire hath devoured the pastures
of the wilderness, and the flame hath burned all the trees of the field. The
beasts of the field cry also unto thee: for the rivers of waters are dried up, and
the fire hath devoured the pastures of the wilderness." Joel 1:19, 20.*

When I lived and worked in China as the ADRA country director in the late 1990s and through mid-2001, I had the privilege of traveling extensively in that country and seeing firsthand not only their huge and sustained economic boom but also the by-product of that boom, their astonishing levels of air, water, and soil pollution. I have vivid memories of rivers and streams so polluted that they stank, seemed to be devoid of life, and were so full of garbage that their surface was just a mass of plastic bottles and other flotsam. Of course, pollution was not confined to surface water. I remember marveling at how thick and how pervasive the air pollution was all through the country; I could board a plane in thick smog, fly two hours with nothing but thick smog below me, and then land in thick smog.

In China there is tremendous pressure on every resource—1.3 billion people in an industrialized economy will have that effect. But there is no resource that is under more pressure than water, and the part of the country that is the most parched is the northern half of the country. Amazingly, much of the problem is human-made through changes China has made as it has literally reengineered its environment.[1] I still clearly remember the moment when I realized just how much China has changed its environment in the past 50 years, and just how much of an impact it has had on the country. I was in an airplane traveling with a distinguished and wealthy gentleman to the north of the country to review potential ADRA projects. The courteous and soft-spoken gentleman was a high-ranking member of China's People's Congress, a very successful businessman, and an Adventist—an unusual mix by any measure. As we sat in the airplane I asked him a few questions about his life, and he filled in the basics, but when I asked him how he had met his wife, he smiled and pointed to the ground, 30,000 feet below us, and said, "I met her here, in the province we are flying over." The air was clear enough that day that I could actually see some of the ground—it was flat land, and I could see the outline of small agricultural fields to the hazy horizon, with only roads, villages, and a few streambeds breaking up the fields.

I sat back and listened as the man told his story. Many years before, during the Cultural Revolution, he had been a college student living in a large city and had been assigned to go to a very rural area and "live with the people, work with the people, and learn from the people." Millions of educated urban dwellers were given the same assignment. The man told me that the villagers had not welcomed him with open arms, and that this was not surprising, since he had no skills they valued and yet he still needed feeding from their resources. His life in the village was very hard—the villagers worked him all day every day and gave him only enough to keep him alive. His only solace was that there was another young college student who had been assigned to the village—a beautiful young woman. Though they were not allowed to interact with each other, the two would occasionally find a way to sneak off at night into the forest, where they would talk, cry together, and strengthen each other. The two supported each other, grew close, and, when the ordeal was over, returned to the city and got married.

When the man finished his story, he sighed. "Those forests gave us cover and saved our lives—I will always remember them."

I smiled at his romantic story and then glanced out the window and grew puzzled. "Didn't you say that the village where you spent the Cultural Revolution was in this province we are flying over?" I asked.

The man looked out the window and nodded. "Yes, we are probably very close to the village now."

I looked out the window again. There was not a tree to be seen below or on the horizon. I turned back to the gentleman. "But there are no forests below us. There is not even a single tree that I can see."

The man nodded. "Yes. It used to be almost all trees, but we needed crops more than trees. Now the trees are gone."

Planet in Distress

When we landed and toured parts of the province that we had flown over, the most pressing need that we saw was for water. A couple decades before and in response to progressive drying conditions, farmers had tried to boost their crop yields by digging wells about 33 feet (10 meters) deep and installing large pumps. The scheme worked fantastically, but the farmers soon found they had not dug the wells deep enough and had to dig deeper and install stronger pumps. After a few years they again found that the water was running out, and they had to dig deeper still. When we visited, the wells were more than 330 feet (100 meters) deep in some areas and could not go much deeper because of cost/technology limits. What, we asked, would the farmers do when their water ran out? They shrugged and said they would move to the city and take any job they could find.

Humans have, in a very significant way, reengineered a large portion of the surface of the earth, spreading out cities, industrial operations, roads, and agriculture all around them. These disruptions in nature are not without consequence on many levels, and one of the most significant consequences is in the availability of fresh, drinkable water. Water storage and availability in the ground is directly related to the health and extent of forests, grasslands, and other "green" areas. These areas not only serve as ground cover, but catch rain when it falls. Plant bases and roots provide percolation routes for water to enter the earth and recharge aquifers. Grasses and other ground cover slow the progression of water over the ground and let it go into the earth instead of just across the surface. Essentially, forests, grasslands, and other green areas are meant to take in water at an irregular rate (such as a thunderstorm), capture much of it, and let it percolate into aquifers, which then discharge it at a regular rate (via streams or springs). Viewed in this way, it is fairly easy to see what happens when natural forests and other ground cover are removed: water flows over the surface of the earth, eroding it and carrying away soil, and much, much less water enters the earth and recharges aquifers. The erosion of soil and a decrease in the amount of water available are linked, as we shall see. Wherever humankind establishes industrial activities or farming or large populations, the surrounding area is reengineered, and, in doing so, the amount of freshwater that recharges local aquifers is usually dramatically reduced. Thus, in most instances, large populations are drawing water from the ground at a much, much faster rate than it is being returned. Just as with bank accounts, constant withdrawing of water will lead to a zero balance.

All over the world, large populations are getting closer and closer to major water crises that will force dramatic changes in the way our societies function. In China extremely large areas were converted from forests and plains into cropland, and they lost the one thing they needed most: water. Now China is engaged in the largest tree-planting effort in the world, but they are starting from a desperate position and are proceeding with very mixed results;[2] it is always much more difficult and costly to restore something than it is to preserve it in the first place.

While the China story is striking, it is not very unusual. Around the world variations of the same story are being played out every day. The seriousness of the situation cannot be overstated. Survivalists (and emergency room doctors) like to quote a maxim: "You can go three minutes without oxygen, three days without water, and three weeks without food." They are a bit in error on the part about food (Jesus went to the absolute human limit when He fasted for 40 days in the desert), but they are right about the first two. The more than 7 billion people on the globe today simply must have water. It is a matter of life and death. And where they do not have it and there are no other options for getting it quickly or cheaply, they will fight to take it from someone else (or fight to get it back from someone who has taken it). The stage is set for significant global conflicts. By 2030 some 3.9 billion people will live in areas of high water stress,[3] and government strategists in affected areas are already preparing for conflict[4] over water resources.

Freshwater availability is imperiled in large swaths around the world not just by our industrial societies and reengineering by human beings, but also through climate change and survival agricultural techniques such as the "slash and burn" method. Worryingly, most of the areas in which freshwater availability is forecast to be most diminished are in the poorest and most undeveloped parts of the world, where the ability to adapt is minimal. When a critical resource like water is in very short supply and there is no real capacity to adapt, there are only two choices for the people who live there: move or die.

About one in three people in the world today already face water shortages.[5] Of these, about 1.2 billion live in areas of actual physical scarcity, while another 1.6 billion face water scarcity because of economic reasons.[6] Water scarcity is defined as a per capita supply of less than 1,700 square meters per year. One square meter of water is a metric ton, 2,200 pounds, or 275 gallons. Thus, water scarcity is defined as less than 1,280 gallons per person per day. Granted, this seems like quite a lot of water for one person each day. However, this amount includes everything a nation uses water for—agriculture, industry, etc.—and 1,280 gallons a day is actually quite a small amount given the demands.

It is possible for a country to have an abundance of water and yet to have large populations in dire need of water. China, for example, is statistically well off in regard to water. But this is misleading. While the south of China has a significant amount of water, the north is chronically dry. Add to this my observation, based on my work and travels in China, that freshwater in China is probably more polluted on average throughout the country than anywhere else in the world, and the net result is that parts of China face incredible water shortages that rival those in Africa and the Middle East.[7] The World Bank has advised China to increase water prices significantly in order to spur conservation, and has warned that China as a whole is on the verge of joining water-stressed countries.[8] Currently, 550 of China's 600 largest cities are running short of water, and the Chinese government even began trucking water to millions of people in the east of the country when wells and rivers ran dry.[9]

But China is not so much a unique situation as it is a forerunner in developing severe water problems. According to a report issued by Resources for the Future, China and all nations that experience freshwater shortages will be forced to explore and implement a variety of adaptation strategies, both from the bottom up and from the top down.[10] Adaptation measures can include such things as switching to composting toilets or capturing storm runoff in city areas and using it for industrial or other suitable applications. However, adaptation inevitably means higher costs and reduced overall water availability. In the United States, water-use laws and adaptation measures are being required by many cities in water-stressed areas.[11] Since the West and Southwest are already short of water and are predicted to get much drier under most climate-change scenarios, these measures are likely to get more widespread and more severe. The problem with cities is the same everywhere: too many people want too much water, and aquifers everywhere are either limited in capacity or are declining.

In the United States it is not just cities that are facing significant water issues. Much of America's agricultural heartland is reliant upon the Ogallala aquifer, a vast yet shallow aquifer that underlies parts of eight states—Texas, New Mexico, Oklahoma, Kansas, Colorado, almost all of Nebraska, and portions of Wyoming and South Dakota. This aquifer provides those living above it with nearly all water needed for agricultural, industrial, and residential purposes. However, for more than 100 years the aquifer has been drawn down as if it were a renewable resource, which it is not, since it recharges very slowly. The aquifer, already shrinking considerably, is now expected to be completely depleted by the year 2035.[12] Any reduction in the water available to America's agricultural heartland will result in decreased food availability and increased food prices. A complete depletion will cause agricultural yields to plummet and send millions of people looking for new homes. And, since the United States exports its grain and other foods on a global basis, the impact of the loss of the aquifer that underlies much of the United States grain belt will be felt around the world. On varying scales and timelines, the same scenario is being seen around the world with other aquifers, particularly in India and Africa.[13] As these aquifers fail, the cumulative impact on food prices and forced human migration will be profound and destabilizing to societies.

Water distribution and availability on the planet are driven by a number of factors but are most fundamentally driven by the hydrologic cycle, where water is evaporated (primarily from oceans but also from land and plants) and spread as rain across the continents. In climate change the hydrologic cycle itself is changing[14] and becoming more intense. This results in more numerous and stronger storms that bring much heavier rain. To the average person, heavier rain means more water being available. However, what matters is not the amount of water that falls on the ground, but the amount of water that actually seeps into it. With floods there is relatively little water that seeps into the earth, contributing to aquifers and raising water tables; instead floods contribute to soil erosion and loss of crops and property and life, especially where trees and brush and other ground cover have been removed. Thus it is that more intense precipitation patterns are contributing to erosion, which in turn decreases soil productivity and decreases vegetative cover, which in turn increases evapotranspiration (a really big word describing the progressive drying of soil though evaporation and plant "breathing"). All of the above conspires to put downward pressure on food production and water availability. In developing countries subsistence farmers are caught in this downward spiral until they give up on the land and human migration is forced because of water shortages.[15] Much of the resulting human migration flows into cities in a push–pull scenario, in which rural people migrating to cities in search of water leads to sharper water shortages in cities.

In addition to precipitation events, increasing global temperatures are leading to changes in the balance between snow and rain and the melting of glaciers. Mountain snowpack volume is critical to recharging aquifers because the relatively slow melting that occurs in summer months allows most if not all of the snowpack to enter the ground and flow into aquifers. Rain, on the other hand, is more likely to run off, causing more erosion and damage as outlined above.[16] Thus, changes in snow patterns to higher levels, such as seen on Mount Kilimanjaro,[17] lead to less groundwater and greater floods with more net damage and, ultimately, less available freshwater.

The same scenario is true of melting glaciers except that, perversely, the melting of glaciers is currently masking the effects of climate change on freshwater availability. More than one sixth of all the people on earth—nearly 1.2 billion people—depend on glaciers and snowpack for their fresh water,[18] and one sixth of all the people on the earth live in the Himalaya region, which is characterized by the inability to store groundwater,[19] thus rendering it almost totally dependent upon glacier and snowpack melting. Currently, the Himalaya valleys and the farmers downstream are enjoying a bountiful water supply because of increased glacier melt. This, however, is like living on credit, and when glaciers melt out, the very many people in the area will likely find themselves without adequate water and, in turn, without adequate food, thus forcing migrations and conflicts of epic proportions. Water shortages also act as a flashpoint for conflict between nations, and such conflict, especially in places such as South Asia, can lead to a broader conflict that potentially draws in other nations.[20]

Where human migration leads to an influx of people into cities, the water problem will continue to plague most émigrés. Many large cities around the world, but particularly those in the developing world, have pressing water shortages,[21] and the issue of water scarcity (and the divide between the rich and poor) is becoming a point of significant civil discontent in so-called megacities around the world. The problem stems from two sources: megacities are often drawing water from surrounding sources at unsustainable rates, and megacities too often skimp on investing in an adequate water infrastructure (or, alternately, quickly grow far beyond the design capacity of the water system installed).[22] The problem of water shortages in major cities leads to yet another problem: rationing of water to farmers and other outlying users so that cities can claim water resources, with a directly resulting impact on food availability and prices.

Freshwater scarcity in coastal megacities (and smaller cities) is also threatened by a push-pull phenomenon in which overpumping of aquifers leads to salt water being drawn into freshwater supplies, resulting in an unusable water supply. Saltwater infiltration into freshwater aquifers can sometimes occur over significant distances. The problem is exacerbated by any rise in sea level, which increases the push of sea water to infiltrate inland.[23] Water with more than 250 milligrams of salt per liter is unfit to drink, as it can cause dehydration. Coastal cities with burgeoning populations face significant risk (and dire results) by overpumping their aquifers, especially as sea levels rise.

And it is not just our underground water that is in short supply. Lake Mead, a source of water for 22 million people in the southwestern United States and which holds a massive amount of water from the Colorado River, has seen water levels in a more-or-less steady decline since 1983 and is in danger of failing as a reservoir.[24] This is a big deal, since Las Vegas and a host of other cities and towns cannot exist without Lake Mead. But it is not just Lake Mead. Reservoirs the world over are in trouble. An example that every Christian will recognize is the Sea of Galilee. Israel relies on the Sea of Galilee for a significant portion of its water, and yet water levels have been falling steadily for some time so that the lake is now nearing the "black line"—the level at which the pumping equipment installed to remove water can no longer function.[25] Without the Sea of Galilee to draw on, Israel is in grave peril, and the conflict that Israel has with its neighbors over the Jordan River[26] becomes a much greater issue that could become a war. Israel's declining reservoir is by no means unique, but it is noteworthy because of the regional history of water disputes and armed conflict over water.[27]

Farmers are especially prone to risk from freshwater scarcity. Climate change is already increasing evapotranspiration in many regions, and precipitation patterns (as already discussed) are trending toward flooding instead of aquifer recharging. To maintain food production, farmers are first forced to rely on wells and then forced to rely on difficult and often expensive conservation techniques. In India and elsewhere farmers in some areas have drawn down aquifers (some of which will recharge only in a geologic time frame) to startlingly low levels. In China water shortages are forcing significant cost and sustainability decisions to be balanced with the extremely important issue of food production.[28] The conflict between water used for farming and water used for industry is especially apparent when considering grain production, in which 1,000 tons of water are required to produce one ton of wheat. That same water, when deployed in industry, can generate 70 times as much productivity.[29] Governments can choose to allocate water to produce food, or can choose to allocate water to produce goods and jobs. In truth, both food and jobs are needed since people need jobs in order to buy food and need food in order to perform jobs. And yet, governments usually favor industrial water allocations over agricultural water allocations.[30] The inevitable result, then, is for food prices to rise so that the water price gap is narrowed. But social consequences of rising food prices are grave and include unrest such as "food riots" and the destabilization of governments, as well as loss of productivity, increased cost through disease, etc. Ultimately, water shortages lead to spirals of societal disruption and decay that serve to increase suffering and decrease food security and social stability.[31]

Taken altogether, it is apparent that the world is in the early stages of a near-global freshwater crisis: Food production will decrease, costs for most products will spiral upward, disease and death will increase, and very large numbers of people will be forced to migrate, causing massive societal tension as well as conflict between nations—war, to use a simpler term. Human beings, who after all were created in the image of God and are intelligent and resourceful, could perhaps solve the problem of a global water crisis if they were united in purpose, if they were not so selfish, and if they did not have any other global problems or conflicts to distract them. However, humanity is not united in the pursuit of any noble or selfless purpose, and humanity has a perfect storm of other problems all coming to a head at the same time—the natural consequence of sin. In such an environment, humanity will of course conceive and implement local and even regional adaptation schemes, but will not be able to undertake any united, concerted actions to cope with the storm that is even now coming upon the world.

As if this were not enough, humanity also has another set of problems—decaying global systems that were created by humans, that are unstable, and that interact with created but decaying global systems. These human-made global systems, which we cover in the next chapter, include our financial and energy (petroleum) systems, and, predictably, both are in real trouble. But not as much trouble as the global complex society human beings have created.

---

[1] Intentional reengineering of the environment and earth is called geoengineering. The Chinese are leading practitioners of geoengineering and hold that it is possible to engineer a lush and productive area where once stood a polluted, overgrazed semidesert. It will take a lifetime of projects before the technique is proven, and it is highly doubtful that we have that long. Until then, it is simply a large-scale, energy-intensive experiment.

[2] Burkhard Bilger, "The Great Oasis," *New Yorker,* Dec. 19 and 26, 2011.

[3] www.oecd.org/dataoecd/29/33/40200582.pdf.

[4] www.commondreams.org/headline/2011/06/30-9.

[5] Climate Institute, www.climate.org/topics/water.html.

[6] "Economic reasons" is a broad category and includes the inability to afford water for sale

as well as because water resources (water infrastructure) have not been developed because of a shortage of funds.

[7] Asian Economic News, "China Facing Severe Water Shortage Crisis, World Bank Says," (January 12, 2009). Available online at www.thefreelibrary.com/China+facing+severe+water-shortage+crisis,+World+Bank+says-a0200795429.

[8] *Ibid.*

[9] John Vidal, "Cost of Water Shortage: Civil Unrest, Mass Migration and Economic Collapse," *The Guardian,* Aug. 16, 2006. See www.guardian.co.uk/environment/2006/Aug/17/water.internationalnews.

[10] A. P. Covich, *Emerging Climate Change Impacts on Freshwater Resources* (Washington, D.C.: Resources for the Future, 2009).

[11] See http://angeles.sierraclub.org/water/Water%20Conservation%20Measures%20Report.pdf.

[12] Katherine Q. Seelye, "Aquifer's Depletion Poses Sweeping Threat," New York *Times* blog, May 4, 2011. Available online at http://green.blogs.nytimes.com/2011/05/04/aquifers-depletion-poses-sweeping-threat/.

[13] Felicity Barringer, "Groundwater Depletion Is Detected From Space," New York *Times,* May 30, 2011. Available online at http://www.nytimes.com/2011/05/31/science/31water.html?pagewanted=all.

[14] Covich.

[15] H. Ketel, *Climate Change and Human Migration* (Paris: UNESCO-EOLSS, 2007).

[16] Climate Institute, *Water.* Retrieved from Climate Institute: http://www.climate.org/topics/water.html.

[17] Steve Connor, "Climate Change Will Melt Snows of Mount Kilimanjaro Within 20 Years," *The Independent,* Nov. 3, 2009. Available at www.independent.co.uk/environment/climate-change/climate-change-will-melt-snows-of-kilimanjaro-within-20-years-1813631.html.

[18] T. Barnett, "Potential Impacts of a Warming Climate on Water Availability in Snow-dominated Regions," *Nature* 438 (Nov. 17, 2005): 303-309.

[19] *Ibid.*

[20] D. Michel, *Troubled Waters—Climate Change, Hydropolitics and Transboundary Resources* (Washington, D.C.: Henry Stimson Center, 2009).

[21] Vidal.

[22] *Ibid.*

[23] P. Gorder, "Climate Change Could Diminish Drinking Water More Than Expected," *Research News,* November 2007. Online at http://researchnews.osu.edu/archive/saltwater.htm.

[24] http://ecocentric.blogs.time.com/2010/10/18/water-lake-mead-is-at-record-low-levels-is-the-southwest-drying-up?

[25] www.israelnationalnews.com/News/News.aspx/148542.

[26] www1.american.edu/ted/ice/Jordan.htm.

[27] www.en.wikipedia.org/wiki/Water_politics_in_the_Jordan_River_basin.

[28] L. Brown, "Falling Water Tables in China May Soon Raise Food Prices Everywhere," *Earth Policy Institute,* May 2, 2000.

[29] *Ibid.*

[30] As an example, see www.sspconline.org/opinion/ANewDimensionofWaterConflict inOrissa_RanjanKPanda_150208.

[31] www.nytimes.com/2011/06/05/science/earth/05harvest.html?pagewanted=all, and www.upiasia.com/Economics/2008/04/22/rice_shortage_threatens_asia/4679.

Chapter 8:

The Decay of Humanity's Energy and Financial Systems and the Fragility of Complex Societies

"The earth mourneth and fadeth away, the world languisheth and fadeth away, the haughty people of the earth do languish. The earth also is defiled under the inhabitants thereof; because they have transgressed the laws, changed the ordinance, broken the everlasting covenant."
Isaiah 24:4, 5.

In the previous four chapters we have reviewed some of God's created global systems and the tremendous decay and corruption that has befallen them since man sinned and Satan usurped control of the earth. Clearly the decay and corruption in these systems is putting greater and greater pressure on human societies around the world—pressure that will continue to increase as systems decay further. But there are other global systems that humans have created that are also subject to decay and corruption through sin and through interaction with decaying created systems. We will look at two of these—humanity's global energy system (petroleum) and global financial system—and then we will look at the peril that humanity faces as a result of building complex societies the world over, and how the decay in natural global systems and human-made global systems magnifies that peril. A great number of books have been written on the two human-made global systems we will review, including some excellent ones in recent years,[1] and thus our review will be a cursory one that skims the surface and gives us just enough information to help prepare for the next chapter, where we try to tie together everything in this and the previous four chapters with the state of the world at present and the nearness of the second coming of Christ.

We will look at just two human-made global systems, though there are more we could include (the global flow of commodities and capitalism itself are obvious choices). However, two are sufficient for our purposes, and we look specifically at these two because, throughout the world, our complex societies rely fully on the functioning of these two systems. In addition, we look specifically at money, because, let's face it, it is the metric almost everyone on earth uses to measure the impact of almost every event or opportunity or threat, including, unfortunately, the impact of the events that precede the second coming of Christ. One of the challenges that we end-of-time Christians face is in learning to disregard the measures the world uses and instead use measures of eternal value—the measure of the value of a soul against the value of our time and resources, for instance. But that is a topic for another time. Back to the subject at hand—humanity's decaying global systems of oil and money—and let's start with our global energy (oil) system.

Oil is the most widely traded commodity in the world. On a global scale, we use more than 32 billion barrels of the stuff a year.[2] Put another way, this means that we pump almost 1.35 trillion gallons of oil each year from deep within the earth, make it do work for us, change it in the process from a liquid to a gas, and then discharge the gas into the atmosphere. You can make a very, very big lake with 1.35 trillion gallons[3] (actually, a body of water that large the Bible would call a "sea") and you can get a tremendous amount of work done with the energy that we harness every year—enough to transform the world into what sinful humanity wants it to be.

By any measure, humanity's use of petroleum is a very big deal that has built global industrial societies, fueled an unprecedented explosion in the availability of food, built and armed vast armies and fleets, and allowed humans to control, to a significant degree, their environment. The use of oil by human beings has also brought about an unprecedented change of the chemical makeup of the atmosphere and seas.[4] However, we've already looked at some of the impact of increased atmospheric CO_2 on the climate. We've also looked at the impact of oil-based pesticides and fertilizers on our food production system, and we've looked at the effect those fertilizers are having on the ocean when they run off the land, so let's move on from the global system impacts of oil and focus instead on what oil has done to humanity and to the societies of humanity the world over.

Oil has given humans power—more power than they have previously had in the history of the world. With the concentrated energy that is in oil and the ability to transform that energy into motion and/or into electricity and thus into all sorts of work, humans are in a much, much better position to turn imaginings into concrete reality than they have ever been. Oil has, to a significant degree, given humans a small amount of godlike powers. And what has sinful humanity done with these powers? A great deal, but in sum humans have used the vast power at their disposal to glorify themselves. We are told: "All defects of character originate in the heart. Pride, vanity, evil temper, and covetousness proceed from the carnal heart unrenewed by the grace of Christ."[5] In the final analysis, human beings have used their extraordinary powers to pursue the desires of their hearts, and have thus separated themselves even further from God. Dwell on that thought for a moment—the net spiritual result of the power humans have derived from oil is a further separation from God.

Human beings were created in the image of the Creator, and thus humans are able to create. With their power, human beings have created vast, highly complex societies all over the world that are dominated by large cities and that are focused primarily on an accumulation and display of wealth and power. Humans have also created an industrial framework that produces a dizzying array of things that humans want or think they need as well as many things that they actually do make use of, such as food and clothing. Humans have, in essence, set themselves up at the top of a pyramid that draws resources from the earth and transforms them into things, which are then used for comfort, for sustenance, and for gratification of ego and pride. For all too many on earth, it is "things" (such as luxury goods) that give human beings a reason to exist. In the context of the history of earth, humanity's folly and vanity is not new. What is striking though is the intensity with which humans have been able to pursue it during the age of oil.

The power humans derive from oil is not limited to remaking the earth and human societies—they have also used their power to expressly reject God's design: in the current day human beings have temporarily overcome the curses God placed upon the earth, to tragic effect. These curses were covered in earlier chapters, but to quickly review, God placed three curses on the earth that each suppressed agricultural production— one when Adam and Eve sinned, one when Cain killed Abel, and the final through the Flood. The proof that humans have in this age temporarily overcome these three curses is simple—how many people reading this book can truthfully say that they have to wrestle with the earth to get their food, that their food comes to them from the sweat of their brow? And in a world in which humans are fundamentally disconnected from the nature that God created, they then are also disconnected from the

Creator. This, then, is a key point—humans have used the power they have derived from oil to insulate themselves to the maximum degree possible from both creation and the Creator.

Human beings have exalted in building complex societies the world over, and these complex societies are underpinned and facilitated through the operation of their global oil system. But humans, even at their best, cannot plan or execute with anything approaching the wisdom of God, and this means that all of humanity's plans are seriously—even fatally—flawed. In the context of humankind's global system of oil supply, and in the context of an increasingly hostile natural world and an increasingly warlike earth, this means that there are inevitable oil supply shortages and disruptions coming, with dire results. In recent history—that is to say, in the past 20 years or so—there has been one distinct example of a country that saw a severe decline in oil imports followed by natural catastrophes that closely resemble those that are expected as our global systems decay. The country is North Korea, a uniquely isolated but still fairly industrialized nation. It is worth considering for a moment the fate of North Korea, since it gives us a clue as to what is in store for highly developed countries that are very dependent upon the continued flow of oil supplies.

North Korea's flow of oil and gas was dramatically reduced when the Soviet Union (their primary supplier and about the only one that would sell to them against the wishes of the rest of the world) collapsed in 1992. The effect was immediate and disastrous. The North Koreans tried to move their energy production to coal and hydropower sources, but they were not able to replace other applications for oil and gas, which were used to make fertilizers and pesticides and were therefore critical. Soon after the loss of oil and gas imports, North Korea happened to suffer epic flooding that washed away topsoil, flooded coal mine shafts, and silted up hydroelectric dams. Complicating this was the fact that the North Korean power and transportation infrastructure began to fail in a domino fashion. As a result, most North Koreans had little coal and no power, and they thus turned to cutting and burning the forests, resulting in deforestation that in turn led to more flooding and lower aquifers. The epic flooding was followed, after a few years, by severe drought. The oil and gas shortage in North Korea had effects on their society that were massive and broad. North Korea could no longer function as a complex and industrialized society, and agriculture output fell by more than 35 percent, resulting in millions of deaths by starvation and a continued suppression of agricultural output.[6]

In the graph below, oil production is forecast. The world's global complex society absolutely needs oil to continue functioning, and demand continues to increase even as production starts to decline. The peak in

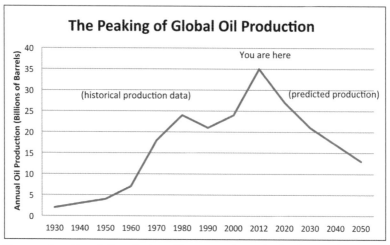

The current peaking of oil production will cause massive societal and economic problems and epic resource conflicts.

world oil production is not a fringe theory: In early 2010 the U.S. Joint Forces Command issued the Joint Operating Environment 2010[7] warning command units of the U.S. military that "by 2012 surplus oil production capacity could entirely disappear, and as early as 2015, the shortfall in output could reach nearly 10 million barrels per day." Further proof of the declining availability of oil is the fact that the commodity has stayed near record price levels during the recent global recession, when industrial demand decreased.

Oil has given humankind amazing power; it has also made humans and their societies extremely weak in the face of a disruption of oil supplies—and a disruption is all but assured in the context of the decay of global systems. It is at this point that the impact of decay in God's global systems and the impact of failure in human's global systems intersect and combine to make a much more significant impact on humans than either set of systems failing on their own. The coming days will not be pretty, and this is especially true for those who live in cities. We have been given, through the Spirit of Prophecy, a great deal of advance information about the state of cities in the last days and the chaos and destruction that will befall them. It is interesting to compare and contrast some of these descriptions with the expected results of disruptions to petroleum flows. If complex societies around the world were to experience interruptions to petroleum flows, it would result, after supplies on hand were used up, in a cessation of deliveries of essential goods and in widespread power outages:[8] no new food supplies, no medical supplies, no material to keep factories running, etc. To the extent that power was not reliant on petroleum supplies, the electricity might stay on, but what good is power without supplies? Imagine a city of, say, 3 million people, in which there is very little transportation available, no food in the stores, a tremendous amount of fear and paranoia everywhere, and, in America at least, a lot of guns on hand. The picture is not a pretty one, and under such circumstances many, if not most, human beings do not often rise to their most noble behavior—the immediate aftermath of Hurricane Katrina in New Orleans is a recent example of this.[9]

We are told of divine judgments coming upon cities, and of disasters of great magnitude coming upon cities.[10] These are different and distinct from the effects of the decay or collapse of human-made global systems such as petroleum distribution, and yet can there be any doubt that the cities struck by terrible calamities—and we are told that "thousands" of cities will be destroyed[11] by such agencies—will leave gaping holes torn in the supply chains that sustain our complex societies, making it all the harder for every other city to survive?

But death and loss and suffering will not be the only result of the disasters attending the days ahead. Another result will be that humanity's insulation from God will be torn away, and in the final days before the close of probation there will be tremendous opportunities for the servants of God to work for Him as people are separated from their luxuries and amusements and are given opportunity to ponder their life and, ultimately, their relationship with God. Given that we can now see these events approaching, now is the time to be planting seeds in preparation for the harvesting of this crop. Writing more than 110 years ago, my favorite writer had this to say: "The plagues of God are already falling upon the earth, sweeping away the most costly structures as if by a breath of fire from heaven. Will not these judgments bring professing Christians to their senses? God permits them to come that the world may take heed, that sinners may be afraid and tremble before Him."[12]

Oil has not made humans evil, but it has given them more power than ever before with which to practice the evil that is in their hearts. And with that power, humanity has constructed something the world has never seen before—a highly complex industrial society that is interlinked at all levels and in all parts of the world so that no part of it is truly isolated from what happens in another part. This is an amazing achievement. However, as calamities and natural disasters increase, as wars and conflicts increase, is it really such a good idea to be so closely linked with and dependent upon all the other parts of the world? What happens to a global society when parts of it begin to collapse? Or, since almost every interaction in our global society is measured or facilitated by money, the question might better be: What happens to humanity's global financial system when parts of it begin to collapse?

Humankind's global financial system is built upon, and is subject to, the health of the world's created global systems. In other words, for our tightly interlinked global financial system to keep working, it is necessary that the world continue to function more or less normally—the climate, crops, water availability, number and intensity of storms, etc., need to stay more or less the same as they have in the past for our global financial systems to remain stable. But, as every Bible-reading Christian knows, in the time of the end—the age we are now in—the world will not function normally at all. The stresses of our current age will place tremendous pressure on the financial systems of the world because they will incur tremendous expense. Check that. They *are incurring* tremendous expense. As cited in chapter 5, the year in which this book was written, 2011, was a record-breaking year in regard to the cost of disasters in the United States. Such costs are more than any society can bear if they come year after year. In this context, what a tremendous pity it is that many Christians in this age spend much of their time and effort chasing money, which will soon enough be worthless to them.

If financial systems around the world were in really good shape, the overwhelming costs associated with the accelerating decay of global systems could wreck those financial systems in the span of not too many years.[13] But around the world financial systems are already in pretty bad shape. At the end of 2011 this is the picture: The United States, the largest economy in the world, is drowning in debt and has no fiscal muscle with which to battle the recession that seems to be winding down. Japan has had a stagnant economy for almost a decade and is reeling from the costs associated with the earthquake, tsunami, and nuclear accident of 2011. The European Union (EU) is staggering from a series of financial blunders and mismanagement, especially on the part of some of its less risk-averse members.[14] Some members of the EU may yet default, throwing the EU into uncharted waters and potentially threatening to break it up.[15] In fact, an EU breakup is being more and more openly considered. China, the world's second-largest economy, is fighting both a recession and inflation at the same time and is also fighting massive corruption[16] and fiscal mismanagement on the part of state-owned companies and real-estate speculators, making the apparent relative health of their financial system quite suspect. And China's housing market—a key economic driver for the past decade—is rapidly weakening, and economic growth is tailing off. At the same time, China is paying the price for monetizing their environmental resources in the last quarter of the previous century[17] and is forced to dedicate increasing amounts of its budget to the alleviation and repair, if possible, of the massive pollution and resource destruction that took place.

Taken together, it is plain that no nation or group of nations has sufficient financial resources to combat the effects of global system decay (including widespread damage to infrastructure, dramatically increased health costs, increased commodity costs, and increased energy, defense, and transportation costs, etc.). Thus, there is no apparent worldly "savior" to look to for a financial rescue from the problems that are breaking upon the world. Worse still, the financial systems of the world are so tightly interlinked that a crisis somewhere in the world inevitably affects the rest of the world. With nations under such pressures in our age, it is extremely likely that we will soon find out what the collapse of a major financial system somewhere in the world does to the rest of the world. And it is extremely likely that such a major financial system collapse will end up increasing pressure on society, with the result that prices increase and thus it will become more difficult for billions of people to live from day to day, which will then massively increase societal pressure and resource competition, which will in turn dramatically increase the likelihood of war.

Thus, as with oil, we find that all societies on earth are tied together by their financial systems, and that, again as with oil, they can neither escape this system nor avoid the shocks and disruptions and conflict that will come when some part of the system collapses. In essence, the highly complex societies that span the globe have been knit into one by oil and money (and other human-made global systems, such as commodities flows). For a time in the history of the earth, this interlinking seemed to make societies stronger. Now the complex societies of the earth are beginning to understand the extent to which this interlinking actually makes them more fragile and vulnerable, but this only places them in an impossible situation because there is no apparent way to be part of the global economy and yet separate from the interlinking that the global economy requires, and no way to withdraw once a country is interlinked.

Planet in Distress

The fragility of complex societies is nothing new. Complex societies have been rising, expanding, and crashing for millennia. But—and this is an important point—this is the first time in the known history of humanity that we essentially have one worldwide complex society.[18] The study of the rise and fall of complex societies is an academic specialty area, and in the field, few are as respected as Joseph Tainter, who in 1988 wrote *The Collapse of Complex Societies,* a widely respected book now held to be a landmark work in the field.[19] According to Tainter the basic definition of a collapsing society is "a rapid, significant loss of an established level of sociopolitical complexity."[20] "Complexity" is commonly used to "refer to such things as the size of a society, the number and distinctiveness of its parts, the variety of specialized roles that it incorporates, the number of distinct social personalities present, and the variety of mechanisms for organizing these into a coherent, functioning whole."[21] Any increase in any one of these factors increases the complexity of a society. While there are many theories about what causes collapse, two prominent ones deal with the depletion of vital resources that a society is dependent upon, and an overinvestment into complexity in an environment of declining marginal returns.[22] However, that technical explanation won't mean much to most people, so let's spend a little time looking into the importance of complex societies and their collapse in search of a less technical explanation that we can use to evaluate the state of our complex global society today and how it fits into global system decay.

Let's start by focusing on a prime aspect of a complex society, which is that some or many members of society are more or less permanently specialized in particular activities and depend, in turn, on others for goods and services within a system regulated by custom and laws. The "specialists" in our societies who do not support themselves through subsistence farming but instead rely on others for basic goods are very recognizable to us: they are soldiers, doctors, nurses, teachers, firefighters, grocers, engineers, computer programmers, businesspeople, writers—in short, just about everyone in most of our complex societies around the world is a specialist. Most people would be very hard pressed to name a society in the world today that is not a complex society. Papua New Guinea has some. Africa still has a few. But that is about it. For the rest of us, the only society we really know is a complex society, which, of course, makes it all the more difficult for us to see it.

We have complex societies because, as populations grow, humans tend to specialize in activities on a full-time basis. These full-time specialists then create demand for specialists in other fields,[23] which in turn require significant amounts of support (in the form of energy and resources) in order to become even more specialized. Specialists tend to create more specialists, and the specialist system tends to demand more and more resources from society.[24] In short, complex societies tend to become very top-heavy prior to collapse, with high taxes on their citizens in an attempt to maintain the levels upon levels of complexity. Sound familiar?

What complex societies tend to do is invest in complexity to solve their problems (we do this now by turning to technology to solve our problems, such as with food production and energy shortages and education and law enforcement and health care and with, well, just about everything that presents itself as a problem). In other words, in our sinful world human beings tend to look to themselves for the solutions to the problems they encounter. From a Christian perspective, this is no surprise. The fatal flaw in this tendency is that, as we continually expand our tiers of specialized activity and invest ever more of society's resources in trying to solve society's problems, we do so at a higher and higher marginal cost (or, in simpler terms, each new solution is more expensive and complicated than the last and makes a smaller impact).

When we do this enough, we eventually undermine our material base and open up obvious opportunities for individual members of society to solve their problems at a lower level of complexity and at a lower cost than through the systems of a complex society. For instance, let's say a country has a food shortage and the proposed solution is to fund university research into hydroponics and then build vast industrial food production facilities using large amounts of energy and large amounts of material and financial resources as well as expensive supplies and highly trained technicians. And let's also say that, given the cost of production, the food produced by this industrial solution will be very expensive. At some point individuals will look at the cost of their food, then look at the soil in their backyard (and front yard) and decide to plant potatoes.[25] They will solve their problem (or at least part of it) using a much less complex and less costly approach. When large groups of people within a society get fed up with bearing the cost of an increasingly complex approach to society's problems and withdraw their support from complexity, the result is either radical change and the discovery of a new fuel for continued expansion of complexity,[26] or society entering the early stages of collapse.

Complex societies that are in advanced stages of resource depletion and spiraling complexity—such as much of the world is today—are under enormous pressure and are much more susceptible to disruptions from external shocks (such as natural disasters) than they would otherwise be. Put another way, societies that are already struggling with an inadequate amount of material and financial resources (as is just about every nation in the world today) can be much more severely impacted by disasters (and shortages, price shocks, etc.) than otherwise, leading to a very unhappy general population and possibly to demonstrations and civil unrest. And here is the critical point: when complex societies are struggling with material and financial shortages, they often turn aggressive and violent. The reason is simple: if further increases in complexity are not yielding returns but the society wants to avoid collapse, then the fastest, cheapest way to acquire more resources is to take them from another, presumably weaker, country.

Let me see if I can put that more succinctly: as the nations of the world are faced with more material and financial shortages (as they are today), and as they face a buildup of societal pressures (as they do today), they are extremely likely to turn to war to solve their problems. They will, of course, dress their actions in flags and patriotic speeches that will stoke the passions of their people, but their core response to their core problem will be to take resources from weaker nations. It is the pathway that is the quickest and that makes the most sense in the logic used by sinful, selfish humans.

The second coping pathway that all nations already use when faced with some external shortage or crisis is adaptation. Humanity is pretty smart, and most societies on earth are able to figure out ways to mitigate or even eliminate isolated challenges by changing how they do things or substituting one material for another. But I assert that humans are not able to adapt and innovate their way out of the cluster of crises that are coming upon the world now. The reason for this is fairly straightforward—humanity is just not able to innovate and adapt rapidly enough to keep ahead of the tsunami of events that are poised to crash upon the world. We have reached a point in our history at which almost every resource the 7 billion people on this earth rely on—oil, arable soil, fish, water—has reached a peak and is in decline. Simply put, the earth is a burning platform and, for those who do not have Christ, there is no way to repair it, though the people of the earth will make desperate attempts to do so while preparing for war.

As adaptation is seen as too expensive, time-consuming, and unsure, and as war prevails as the survival pathway of choice, Satan will step in posing as a problem solver ("savior") for humanity. And humanity will be deluded by Satan in this guise. A passage in *The Great Controversy* puts it like this: "While appearing to the children of men as a great physician who can heal all their maladies, he will bring disease and disaster, until populous cities are reduced to ruin and desolation."[27] The bottom line is that humans, who have relied upon their own talents and who have grown away from God over the history of the earth, will not be able to solve or even partially address the overwhelming problems that are even now breaking upon us. In the final days—in our age—a few will rely upon God, and the balance will rely upon Satan, to their utter ruin.

When you combine the weaknesses of the two human-made global systems with the teetering state of humanity's global complex society, and then combine these with the decaying, deformed, and dying global systems that the earth relies on to operate, it is obvious that there are grave troubles ahead. And when you factor in our inability to adapt effectively, it is obvious that the world is slipping rapidly into a state such as has not been seen in the history of humanity. The more we learn about the precarious position of the 7 billion people who live in the world today, the more we see that humanity is about to be overwhelmed by events. Only those who cling to Christ will make it through the storm.

---

[1] Including *The Party's Over, Powerdown,* and *The Oil Depletion Protocol,* all by Richard Heinberg, as well as *The Long Emergency,* by James Howard Kunstler, and a must-read, *Twilight in the Desert: The Coming Saudi Oil Shock,* by Matthew Simmons. On global financial systems, *Crisis Economics,* by Nouriel Roubini and Stephen Mihm, is excellent, and *Return of Depression Economics,* by Paul Krugman, is a must-read. An excellent book to read as an overview is *Peak Everything,* by Richard Heinberg.

[2] www.cia.gov/library/publications/the-world-factbook/rankorder/2174rank.html.

[3] That many gallons, if it were water, would be about one fourth of all the water that flows through the Colorado River in a year.

[4] The chemical change in the seas has happened because the seas absorb excess gases from the atmosphere. In this case they have absorbed enough CO_2 to noticeably change the acidity of the seas and to threaten life.

[5] Ellen G. White, *God's Amazing Grace* (Washington, D.C: Review and Herald Pub. Assn., 1973), p. 223.

[6] www.news.bbc.co.uk/2/hi/asia-pacific/433641.stm. Much of the background on North Korea used here was derived from Dale Allen Pfeiffer, *Eating Fossil Fuels,* a highly recommended book.

[7] See "Peak Oil," *Wikipedia,* for a pdf of this document. http://en.wikipedia.org/wiki/Peak_oil.

[8] On a global basis, oil is still widely used to generate electricity, though the use of oil for this purpose in the United States has declined significantly as it has been replaced with cheaper, cleaner natural gas.

[9] See David Brunsma, David Overfelt, and Steve Picou, *The Sociology of Katrina: Perspectives on a Modern Catastrophe* (New York: Romon and Littlefield, 2007).

[10] See E. G. White, *Country Living,* for an overview of some of her statements on the subject.

[11] Ellen G. White, *Evangelism* (Washington, D.C.: Review and Herald Pub. Assn., 1946), p. 29.

[12] Ellen G. White, *Manuscript Releases* (Washington, D.C.: Ellen G. White Estate, 1981-1993), vol. 3, p. 311.

[13] Nobody really knows how much the wide array of impacts related to global system decline will "cost" us, partly because we have an imprecise measure of just how bad things really are, partly because we don't really know how to calculate the "cost" of a world in steep decline, and partly because we don't really want to know.

[14] Greece and Italy are the most prominent among these, but there are other less-than-responsible members including Spain and Portugal.

[15] See www.en.wikipedia.org/wiki/Seventh-day_Adventist_eschatology. Also see www.guardian.co.uk/world/2011/nov/03/g20-freece-eurozone-crisis-mood?newsfeed=true.

[16] See www.piemonte.indymedia.org/article/5391 and www.eastasiaforum.org/2011/05/10/mega-population-mega-corruption-mega-growth/.

[17] In other words, they turned their natural resources into money and destroyed their natural resources with pollution in order to keep production costs low—now they are paying the price and must restore their "natural capital" at costs that are magnitudes of order above the values of the resources they monetized.

[18] The world may have had a global complex society prior to the Flood—we simply do not know.

[19] Those interested might want to read *Collapse,* by Jared Diamond, the same scholar who wrote the highly acclaimed *Guns, Germs, and Steel.* Diamond is somewhat more accessible than Tainter, who writes for an academic audience.

[20] Joseph Tainter, *The Collapse of Complex Societies* (Cambridge: Cambridge University Press, 1988), p. 4.

[21] *Ibid.,* p. 23.

[22] *Ibid.,* pp. 42, 118, 119.

[23] Just as a doctor creates demand for nurses, who then both eventually create a need for a pharmacist, etc.

[24] Universities as systems and the cost of a university education are perhaps perfect examples of this phenomena.

[25] Potatoes are the most energy-dense crop it is possible to grow on a square-yard basis. They do not provide complete nutrition, but if all you have is a limited space, you will get much more to eat by planting potatoes than by planting, say, lettuce or wheat or squash.

[26] A good example of this "forced innovation" would be the start of widespread use of coal after much of England was deforested in the medieval era.

[27] E. G. White, *The Great Controversy,* p. 589.

Chapter 9:

Global System Decay in the Context of the Great Controversy and the Time of Trouble

"Lift up your eyes to the heavens, and look upon the earth beneath: for the heavens shall vanish away like smoke, and the earth shall wax old like a garment, and they that dwell therein shall die in like manner: but my salvation shall be for ever, and my righteousness shall not be abolished."
Isaiah 51:6.

In Mark 13 much of what we are told in Matthew 24 is repeated, but the chapter ends differently—the final verses (35-37) say: "Watch ye therefore: for ye know not when the master of the house cometh, at even, or at midnight, or at the cockcrowing, or in the morning: Lest coming suddenly he find you sleeping. And what I say unto you I say unto all, Watch."

And overall, we Adventists have watched, but not all of us have watched the same things, nor, I suggest, *have* we been quick to see the big picture of events unfolding around us. During my lifetime every president of the United States and every pope of the Catholic Church has been suggested as being the antichrist by some person or group on the fringe who was watching. Mikhail Gorbachev was also nominated to the position on the strength of being a Russian and having a birthmark near his forehead. More within the mainstream, during my lifetime every social upheaval (hippies, drugs, immigrants, Wall Street crashes, oil price spikes) has been pointed to as a specific sign of Christ's soon coming, and of course natural disasters and such phenomena as large volcanic eruptions have received prominent mention. Over the years Adventists who tried to "connect the dots" between current events and biblical prophecies were, to one degree or another, heeding the counsel of Mark 13. They were also being very human, for it is basic human nature to look for a big, single, situation-changing event when "watching."

The truth, though, is that we have been so busy looking for something big and striking that we may have missed the fact that the weight of sin on the earth grows heavier every day and that the world is steadily, daily, grinding toward a painful close. Could it be that most of us have missed the accumulation of daily, very small events that, in aggregate, point with some clarity to the soon coming of our Savior? Such an error could well be a fatal error for some, as the lack of a big single event pointing to Christ makes it easier for some to drift toward the world and drift along with the world, eventually becoming so enmeshed in the world that they miss the soon-coming and more obvious signs of Christ's return. In this context there is much to learn for our day in the parable of the 10 virgins and the bridegroom that tarried in Matthew 25.

It is ironic that so many non-Adventists and non-Christians are watching the signs and forces at work in the world and are amazed (and on the verge of panic) at what is coming on the world as a result of global system decay, resource shortages, population pressures, etc. It is again ironic that it is sin that is at the root of these problems, and that only Christians are able to provide the world with some explanation of what is going on, and that only Adventists are able to provide a clear explanation in the context of the great controversy between Christ and Satan. How sad it is that people of the world see the signs of momentous events in the decay of God's creation and that they see the imminent conflicts the world faces while Adventists are either mostly mute on these points or have their attention focused elsewhere.

The good news, though, is that we can still wake up—and wake up we must. The even better news is that very soon—imminently, in my opinion—the latter rain will be poured out abundantly, and those who are not asleep but who are working in the Lord will work with power. It behooves us, then, to heed the advice of Mark and look carefully not just at individuals or single events but at the whole pattern of events in the world and compare these with what our Lord said was coming just before the time of trouble. It also behooves us to repeat the same exercise comparing the pattern of events and trends in the world to what we read in the Spirit of Prophecy.

The preceding chapters have given us a clearer picture of the current and expected effects of global system decay not only on nature but on our global economies and on our global complex society. With this in mind, let's turn to the Bible to review the state of the world just before Jesus comes. Let's turn to Matthew 24:6, 7 to read the words of Christ as He describes the signs of His coming at the close of earth's history: "And ye shall hear of wars and rumors of wars: see that ye be not troubled: for all these things must come to pass, but the end is not yet. For nation shall rise against nation, and kingdom against kingdom: and there shall be famines, and pestilences, and earthquakes in divers places."

Let's take those verses apart piece by piece and look at each carefully in the context of what we have been studying in this book. The following table will help us do that by comparing passages in the text to expected impacts of global system decay.

Passage	Import in the Context of Global System Decay
"And ye shall hear of wars and rumors of wars:"	There is building pressure for war from many directions (food shortages, finances, energy shortages, weakness from disasters, etc.), which together push or pull nations to conflict.
"see that ye be not troubled: for all these things must come to pass,"	Disaster and death are the natural end of Satan's administration of earth. All of his claims are disproved conclusively by the evidence of the state of the earth and its people. These things must happen to close the great controversy for all time and thus assure that sin does not reoccur anywhere in the universe.
"but the end is not yet."	The testing and proving of the time of trouble and close of probation must yet take place before Christ returns to claim His faithful.
"For nation shall rise against nation, and kingdom against kingdom:"	The world stage is set for conflict: complex societies turn to violence to try to increase resources or decrease threats; nations try to solve their resource shortages through conflict.
"and there shall be famines,"	The world is experiencing significant hunger now, and the stage is set for much more severe hunger that will threaten the lives of millions upon millions. Almost every impact of global system decay reduces agricultural output, while wars cut off supplies and further reduce production.
"and pestilences,"	Pestilences have already increased, and science predicts that pestilences will be a key component of climate-change scenarios.
"and earthquakes in divers places."	Satan cannot maintain what God created, and God's earth will continue to decay and grow more and more unstable, including the very foundations of the earth.

Now let's look at chapter 36 of *The Great Controversy* (which was written more than 100 years ago) and see what is predicted to come upon the earth just before the time of trouble:

From *The Great Controversy*, Chapter 36	Import in the Context of Global System Decay and Scripture
Satan delights in war	Matches perfectly with Matthew 24, and with expectations of resource-based conflict consistent with the precollapse state of complex societies.
Satan will bring disease and disaster, until populous cities are reduced to ruin and desolation	Fits perfectly with the concept of complex and interdependent societies being very fragile. Fits also with the concept of population centers being especially vulnerable to the expected impacts of global system decay.
accidents and calamities by sea and by land	Fits perfectly with climate-change projections.
great conflagrations	See Texas, 2011.
fierce tornadoes, terrific hailstorms	Fits with climate-change projections. Also see southeast U.S., 2011
tempests, floods, cyclones	Fits with climate-change projections. Also see northeast U.S., 2011; India, 2011; Central America, 2011, and elsewhere.
tidal waves	See Japan, 2011.
earthquakes	See Washington, D.C., 2011; also Japan.
famine and distress	Fits with climate-change projections. See Horn of Africa, 2011.
deadly air and pestilence	Fits with climate-change projections. See China, 1994-2011; also India.

Looking at the two tables, two things seem obvious: First, that the events around us confirm that the last days are upon us. Second, that the current level of intensity of events is relatively low and we can expect them to build in intensity considerably, especially in regard to war and the destruction of cities. An objective observer could claim that any similar charts made in any year since Ellen White penned *The Great Controversy* could make a similar claim. Fair enough, for there have been disasters every year since that time and for millennia stretching back through recorded history. But there are two reasons today is distinctly different. One of these is that natural disasters are increasing both in intensity and frequency,[1] though this is difficult to prove, since we have only been able to observe and record the globe on a

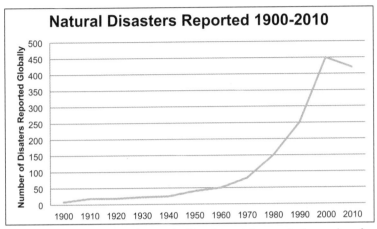

Progressively better reporting has resulted in a dramatic increase in the number of disasters reported, thus skewing the results in the chart above. Still, there seems no doubt that disasters are increasing significantly, especially since 1980.
Source: www.emdat.be/natural-disasters-trends

comprehensive basis for the span of a few decades, and thus better observation certainly accounts for part of the rise in the graph on this page. The primary reason, though, is this: for the first time since Matthew recorded the words of Christ and for the first time since Ellen White penned *The Great Controversy* we can actually see and measure and monitor the forces that are building and inexorably fulfilling the predictions regarding conditions that will ripen in our time—in these days just before the time of trouble.

If the last days before the time of trouble were to be compared to a storm, then from the time of Christ we have been told about the storm that is coming. And from the time the Adventist Church started, from the pioneers down to our day, we have been expecting the storm and have been watching for it. Some have said they saw distant clouds on the horizon. But now, in our day, the storm is not only on the horizon, but we are able to see the dark clouds build and billow, and we are able to hear the thunder and see the flashes of lightning with the clouds. We are able to feel the building wind and are able to smell the storm within that wind. We are even able to feel the first few light drops of rain and are reminded that this same storm will bring with it an outpouring of the Holy Spirit. From this we know that the days left in which we can work are few, but we know that in these final days we will work with power.

While those who watch and seek God in His Word know that *the* storm is approaching, the political leaders of the world are aware only that *a* storm is approaching and are actively preparing for looming resource shortages and for conflict. The forces at work that will bring about the predictions of Matthew 24 are clearly apparent to them, though they do not think of them in spiritual terms but in terms of wealth, power, and control. Scientists and thought leaders from many backgrounds are recording and interpreting data and comparing their findings to the findings of others, and as a result they are publishing papers that are the academic equivalent of stern, dire, and even panicky warnings. They too clearly see the forces at work that will bring about the predictions of Matthew 24, even if they dismiss the Bible as an old book full of fables.

The challenge to Adventists in these days is to loudly proclaim Christ to a world focused on wealth, power, control, and gratification, even though we know that proclaiming Christ will make us a target for ridicule by friends and neighbors (and, in some cases, even by current or former church members). This is the time to fully and finally abandon the world, stop caring about reputation or pride or status or gain, and unabashedly proclaim Christ, all the while expecting no better reception than Christ Himself received.

We have numerous biblical and extrabiblical accounts of the events that are coming,[2] but in using these it is important to make a strict distinction between the events before the time of trouble and the events during the time of trouble. Such a distinction is important within the context of this book because the actual events of the time of trouble will be precipitated by God removing His limitations on Satan while He at the same time ceases to hold back the winds of strife. At the same time, God will pour out His wrath upon the earth. This direct action is very different from the consequences of sin seen in nature, and it is important not to confuse the two.

To be clear: It is the results of sin, as seen in the decay in the earth and in the character of man, that will bring the world to the state that Christ predicted. But they will not precipitate the sudden destruction and chaos that come in the time of trouble and especially after the close of probation. There is a distinct difference between sin-induced decay and things that happen through the hand of God. After the four angels cease to hold back the winds, both God and Satan will take a much more direct role in the final events of the earth. No description of the days that are upon us, or the order of events, is clearer than this passage in *Maranatha,* penned more than 100 years ago:

"While the work of salvation is closing, trouble will be coming on the earth, and the nations will be angry, yet held in check so as not to prevent the work of the third angel. At that time the 'latter rain' or refreshing from the presence of the Lord, will come, to give power to the loud voice of the third angel, and prepare the saints to stand in the period when the seven last plagues shall be poured out.

"I was shown the inhabitants of the earth in the utmost confusion. War, bloodshed, privation, want, famine, and pestilence were abroad in the land. As these things surrounded God's people, they began to press together, and to cast aside their little difficulties. Self-dignity no longer controlled them; deep humility took its place. Suffering, perplexity, and privation caused reason to resume its throne, and the passionate and unreasonable man became sane, and acted with discretion and wisdom.

"My attention was then called from the scene. There seemed to be a little time of peace. Once more the inhabitants of the earth were presented before me; and again everything was in the utmost confusion. Strife, war, and bloodshed, with famine and pestilence, raged everywhere. Other nations were engaged in this war and confusion. War caused famine. Want and bloodshed caused pestilence. And then men's hearts failed them for fear, 'and for looking after those things which are coming on the earth.'

"Angels are now restraining the winds of strife, until the world shall be warned of its coming doom; but a storm is gathering, ready to burst upon the earth, and when God shall bid His angels loose the winds, there will be such a scene of strife as no pen can picture."[3]

We live in a day when we, as Seventh-day Adventist Christians, have more knowledge of coming events than any people have ever before possessed. We have from the Bible a broad account of what is coming and what we are charged with doing. We have from the Spirit of Prophecy a detailed account of the events of the days that are now upon us. And now we have from science and from thought leaders around the world an analysis of the forces that are bringing us to the state of the world as described by our Savior in Matthew 24. And with the fullness of this knowledge, it is possible for the discerning to track the gathering forces of destruction and disaster, and to measure the building conflicts and waxing forces of war around the world.

Here then, brothers and sisters, is the key summation of everything that has been covered in this book up to this point: Christ really is coming, and He really is coming soon. I am convinced He will come in our lifetimes. Full stop. And with this great knowledge comes great responsibility—what will we do to prepare ourselves, our families, our friends, our neighbors, and our world for the imminent coming of Christ?

---

[1] www.emdat.be/natural-disasters-trends.

[2] An excellent collection is available at the E. G. White Estate Web site: www.whiteestate. org/issues/Conflicts.asp.

[3] E. G. White, *Maranatha,* p. 259.

Chapter 10:

Preparation and a Ministry for Everyone

"We are living in the most solemn period of this world's history. The destiny of earth's teeming multitudes is about to be decided. Our own future well-being and also the salvation of other souls depend upon the course we now pursue."—Ellen G. White, The Great Controversy, p. 601.

"Wherefore He sayeth, Awake thou that sleepest, and arise from the dead, and Christ shall give thee light. See then that ye walk circumspectly, not as fools, but as wise, redeeming the time, because the days are evil." Ephesians 5:14-16.

Jesus really is coming. But before that, Sunday laws really are coming, and the latter rain really is coming. Tribulation and the time of trouble really are coming, the close of probation really is coming, and yes, the final trying events and the days in which we must stand before the world really are coming. And all these things are coming very soon—I sincerely believe that Christ is coming within my lifetime, and I am 51 years old at the writing of this book. Brothers and sisters, there never was a more urgent time to prepare ourselves and our families for both the opportunities and trials that are immediately ahead.

There are only a few things that we can do that will allow us to minimize the impact of the global trauma that is coming. There is, however, a great deal of work that we must be engaged in that will help others to have eternal life. And at the same time, there is much that we must do to strengthen ourselves for the trials ahead by growing to truly know Christ and developing our personal friendship with Him. In this context our thoughts and energies should be directed toward Christ, who is our only protection, toward the study of His Word, which is our only refuge, and toward the faithful execution of the duties He has assigned us in these, the closing days of history.

In this chapter we will review the things we can do to minimize the impact on us of the global trauma that is coming, and we will look at what preparation each of us needs to make as individuals to meet our Savior, as well as what we can do to help our family, friends, and community prepare to meet Jesus. And we will look at interesting approaches to ministry that anticipate the needs created by global system decay and use them as an opportunity to witness.

We know from our review of the decay of global systems that the natural world is beginning to fall apart and is certainly becoming less productive and more hostile. We also know that humanity's global systems—oil and money—are in decline and subject to shocks and crises. Finally, we know that humanity's global complex society is fully dependent upon the smooth functioning of both natural and human-made global systems, and we therefore can conclude that society the world over will be convulsed and shaken with a multitude of crises, shortages, disasters, and wars. Within this context our course is simple: decrease our dependency upon the world and "withdraw [our] affections from the world, and fasten them upon Christ."[1] Here are four specific ways in which we can decrease our dependency upon the world:

- **Eliminate debt.** When we are in debt to the world, we are beholden to the world, and we must remain engaged in the world in order to maintain our debt payments. In order to freely serve the Lord whenever and wherever He calls, we urgently need to eliminate debt. Doing so will likely require that we radically change our lifestyles in favor of a simpler existence. It may also require that we part with some or much of our material goods. In view of what is coming, such changes may be exactly what the Lord wants us to make in our lives. By seeking the Lord in prayer and by seeking His counsel in the writings of Christian authors,[2] we will know what to do in regard to getting out from under the burden of debt.

- **Leave the cities and suburbs.** Adventists, especially those with children, have long been urged to leave the cities.[3] How much more urgent is this advice in our day when the spiritual condition of humanity has continued to decline? On a practical level, our study of global system decay has taught us that cities are extremely vulnerable and fragile and that those living in them will suffer greatly in the coming days. This view is reinforced by inspired writings, in which we are told that cities will suffer terribly in the coming days: "The end is near and every city is to be turned upside down every way. There will be confusion in every city. Everything that can be shaken is to be shaken, and we do not know what will come next."[4] We have been counseled to leave the cities for less-populated areas and for country areas, and have been further counseled to remain prepared to leave less-populated areas for remote mountain areas when Sunday laws are passed.[5] Those who have not taken the first step should in solemnity lay this before the Lord and seek His will; all should begin to give thought to taking the second step in the days ahead.

- *Plant a garden.* It is important to note that when we were counseled to plant gardens and fruit trees, it was so that we could have healthy food, and we were told that the land was to provide our necessities.[6] We have also been counseled that being close to nature by gardening would bring us closer to God. What a wonderful combination we find in becoming less reliant upon the world while at the same time we are drawn closer to God. For those who have space for a garden but who do not have a garden, it may be best to start a modest garden, gaining experience and confidence with raising, preparing, and preserving your own food. The garden space can be increased—even doubled—every year until it reaches a size that is manageable while at the same time producing significant amounts of food.

- *Simplify our lives.* The suggestion that we simplify our lives comes directly from the Spirit of Prophecy. But it is very interesting to juxtapose that advice with the fact that we live in a global complex society. Almost by definition, if we withdraw from a complex society we will be required to simplify our lives. We will do without entertainments produced by people who specialize in making them. We will do without luxurious foods and clothing, we will decrease our energy consumption, and we will either produce more of our own goods or will do without. Simplification also involves rejecting the enticing snares of Satan, such as television or access to worldly diversions or inappropriate content via the Internet. One of the things that occurs when we simplify our lives is that we create time and space in our lives by rejecting worldly activities. To the extent that we then fill this void with activities that bring us closer to Christ,[7] we will receive incredible spiritual blessings.

Everything on the above list is extremely important and will require significant—perhaps extreme—effort on the part of those who undertake them. At the same time, accomplishing some or all of the things on the list will help us make a significant separation from the world, and in the days that are coming this separation will be a massive blessing to us.

The types of preparation listed above are, however, of secondary importance compared to preparing our hearts and minds to meet our Savior and, in the process, truly coming to know Him as our friend. There are an abundance of materials available from both Adventist and non-Adventist sources that can be used to help guide us closer to a personal relationship with Christ; four key steps are covered here. Adventist author, speaker, and pastor Lee Venden likes to say that, in regard to getting into heaven, "it is not *what* you know but *who* you know." Our first priority, then, in preparing for the second coming of Christ is to cultivate and deepen our personal relationship with Christ. Perhaps the best place to start is with the relatively short book *Steps to Christ*. I can also personally recommend Lee Venden's CD series *All About Jesus*.

The preparation of our hearts and minds is, in part, so that we may witness to the world. In this context it is essential that we not confuse decreasing our dependency on the earth with withdrawing ourselves from the people of the earth. Indeed, Christ said: "Ye are the salt of the earth" (Matt. 5:13). And in *Humble Hero* we read: "Do not withdraw yourselves from the world in order to escape persecution. You are to live among people, so that the distinctive quality of divine love may be like salt to preserve the world from corruption."[8]

The first step, then, is to try to see in each person—everyone we come in contact with—a precious soul for whom Christ died and to see in this soul an opportunity for us to share Christ's love and compassion, an opportunity to share our encouragement, and an opportunity to share our hope. We as Christians need to develop a set of metrics that are completely different from what the world uses. We need to come to the point where "value" and "wealth" do not apply to money or precious metal or material resources, but instead apply to souls and friendships and opportunities to share encouragement, faith, and Christ's love. If we do this, we will better know Christ by valuing what He values.

The next step in our personal preparation is to study the Bible and store it up in our hearts, specifically seeking to know and understand Christ through the Scriptures. We read such advice very clearly in *Prophets and Kings*, in *The Great Controversy*, and in *Testimonies for the Church*. In *Prophets and Kings* we read: "Christians should be preparing for what is soon to break upon the world as an overwhelming surprise, and this preparation they should make by diligently studying the Word of God and striving to conform their lives to its precepts."[9] In *The Great Controversy* we read: "None but those who have fortified the mind with the truths of the Bible will stand through the last great conflict."[10] Finally, in regard to regular daily study and memorization of the Bible, we read in *Testimonies for the Church:* "Several times each day precious, golden moments should be consecrated to prayer and the study of the Scriptures, if it is only to commit a text to memory, that spiritual life may exist in the soul."[11]

In conjunction with placing the Bible in our hearts, the third step in preparing to meet Christ and preparing for the time ahead is by developing a vigorous prayer life. This development of a communion with God is a critical step. In *The Ministry of Healing* we read: "Communion with God will ennoble the character and the life. Men will take knowledge of us, as of the first disciples, that we have been with Jesus. This will impart to the worker a power that nothing else can give. Of this power he must not allow himself to be deprived. We must live a twofold life—a life of thought and action, of silent prayer and earnest work."[12] Note that we must not stop at prayer. Indeed, in *Testimonies for the Church* we are told: "Prayer and effort, effort and prayer, will be the business of your life. You must pray as though the efficiency and praise were all due to God, and labor as though duty were all your own."[13] We are also told: "He who does nothing but pray will soon cease to pray."[14] And so we see that action is necessary, not just in addition to prayer, but in addition to undertaking all of the steps above. Further, we should not make the mistake of waiting to take action until we think we have reached an advanced state in all of the above steps—action and work on our relationship with Christ are both more productive when they are taken simultaneously.

The fourth step that we should take in preparing ourselves to meet our Savior is to study and understand the events that will unfold in the last days, so that we are prepared for them and so that we understand the nearness of His coming. We will be less likely to be swayed by news or fears or various crises if we have an understanding, grounded in Scripture and the Spirit of Prophecy, of the overarching events of the last days. Adventists have a wealth of resources to help with such a study, and *Last Day Events,* by Ellen G. White, is perhaps the best place to start.

After you are fully involved in the task of preparing yourself for meeting your Savior, the next set of activities to undertake is the preparation of your family for the final days. One of the most effective ways to help shift the focus in your family to spiritual things is to withdraw from the world, and, conveniently, some of these steps are similar to the things you can do to minimize the impacts of the global trauma that is coming. You can start preparing your family to meet their Savior by decreasing their dependency on the world and becoming more self-sufficient. As previously stated, this definitely means getting out of the city and growing and preserving some or much of your own food. But it also means learning much, much more about natural healing methods and therapies as outlined in *The Ministry of Healing* and as covered by an array of Adventist sources.

In addition, you can also distance yourself from the world by removing the media connections that incessantly speak for and promote the world. You can make a concerted effort to rid your home of the channels through which the enticements of the world may enter, including television and radio and magazines with superficial and worldly content. And you can further decrease dependency upon the world by choosing to practice simplicity (which also helps eliminate debt as quickly as possible). There is an abundance of additional guidance available in writings published by the Adventist Church, and perhaps the best place to start is the short book *Country Living*. Of course, all of the above assumes that your family *wants* to undertake preparation for the final days. If this is not the case, then you have an opportunity to show them Christ by growing close to Christ and letting your family see Him through you. Whatever course you choose, be aware that Satan will do everything possible to defeat your efforts and demoralize you—in all of your witnessing you can expect one of the biggest battles of your life, and you can expect that only your Friend and Savior Christ will win the battle for you.

There is much that you can do to prepare your friends and neighbors for meeting Christ, and happily, much of it is an extension of what you can do to prepare yourself. In fact, the first and best thing you can do is to draw so close to Christ that others see Him in you. The next thing you can do is to hold them up in prayer and ask God to create opportunities for you to witness to them. You can and should also give of yourself to them and interact frequently with them, engaging them in discussion and working alongside them, seeking opportunities to perform an act of kindness—we are far less effective if we try to witness from afar and fail to show that we are truly interested in the people to whom we are trying to witness. When we get to know people, we will also get to know their challenges and hurdles and thus have opportunities to offer to pray with them and present their problems before the Lord. Finally, we should have a store of books and articles on hand that we have read and that we can recommend as relevant to the challenges our friends and neighbors face.

When it comes to serving our community (which includes the entire world), the bottom line is very simple—everyone should have a ministry. Whether large or small, celebrated or virtually unknown, everyone should have a special area of effort and focus that they are making for their Savior. Our age is not the time for idling or for making half efforts—we must be about our Father's business. Certainly there are no shortages of opportunities to help people, and each person should seek the will of God in deciding on the ministry that best suits their talents and resources. That said, here are some ministry ideas that are built around the needs that will very likely be precipitated by the continuing decay of global systems:

- The worldwide food price spiral and food shortages that are expected as a result of global system decay will create a multitude of personal ministry opportunities, including opportunities to conduct a "greenhouse ministry" whereby those with even a modest little greenhouse can get plants started in their greenhouses (or even their windows) and then offer the plants (plus assistance) to people who want and need to convert some or all of their backyard into gardens but do not know how. Better yet is a ministry that helps those in need understand how to garden and prepare to garden, perhaps even by offering to help them rototill their land (a "rototiller ministry"—whoever thought those two words would go together?). By offering plants and offering of our time and tools and knowledge, we can help people in a concrete way while at the same time forming a relationship with them that will allow us to witness effectively. Another advantage in forming a "greenhouse ministry" is that there are reasons throughout the year to visit new gardeners (pests, fertilizer, pollination, thinning, etc.) and thus opportunities to maintain and expand the relationship. If several people in a church are joined together in a "greenhouse ministry," then the facilities of the church should be used to hold a "food preservation seminar" that includes blanching, freezing, canning, drying, etc. This will extend opportunities for witnessing and for extending relationships, and will also make it easier to invite new gardeners back to the church for a worship service.

- One of the effects of global system decay is that prices for everything will rise while opportunities to generate additional income will decline. This will mean that everyone will be looking for opportunities to decrease costs. In the United States, one of the areas in which costs have steadily spiraled for decades is in health care. Health-care costs have risen to the point that many Americans are already forced to make extremely difficult decisions on whether they will pay for health-care services or for food or rent. In the context of the coming days and expected rising costs and decreasing incomes, it is apparent that more and more people will be forced to seek alternatives to the high-tech medical services that now prevail in America. This dynamic leads to robust ministry opportunities for those who want to help people by offering simple, home-based treatments that involve diet, massage, hydrotherapy, and other interventions described in such books as *The Ministry of Healing* and *Counsels on Health*. Further, by showing true care for people and by helping them in a concrete way, we will open up opportunities for witnessing that would not otherwise be available to us.

• Increased global catastrophes as already experienced are expected to rise dramatically under global system decay, and this means that people the world over will be suffering and in need. Such disasters create numerous opportunities to lend assistance, whether by lending financial and prayer support to ADRA, or by participating in short-term missions and relief trips in conjunction with ADRA or in some other church-sponsored context. It is impossible to go on one of these trips and not be touched and changed for the better by the people that you meet. This being the case, such trips are an excellent "in-reach" to new Christians coming into the church and to longer-term members who may have drifted to the periphery of church and spiritual life.

• As the world becomes more and more unstable, people everywhere will have one question on their mind: What is going on? For those who are called to an information ministry, this question will create opportunities for dialogue and witnessing that would not otherwise exist. Whether through a local door-to-door ministry that involves handing out church literature or selling church books, or through an Internet-based ministry that attempts to answer people's questions and point them toward the Bible and Christ, or through a writing ministry that provides people with insights and information and motivating thoughts—all of these are needed. Some Adventist churches are having very significant success with their own local FM radio stations,[15] finding that they can carry a service such as 3ABN radio most of the time and offer their own programming, such as the church service and educational seminars or Revelation seminars, as often as is possible. Such local programming could extend to the gardening and health seminars suggested above. A local FM radio station that uses prepackaged programming most of the time reduces the burden on the church while still promoting "local" material that spurs visitors to the church.

- Finally, the resource scarcity and economic strains of the coming days will create a large number of people who are economically distressed, perhaps jobless, and in need of assistance. The community service centers of our churches can become critical resources and focal points for both service and evangelism. Clothing donations and food pantries will become far more critical than they have been in the past. Day-care ministries may provide a self-supporting or partially self-supporting opportunity to witness to children and parents while allowing desperate parents (often single parents) to have or keep a job. What could be more important and more fulfilling than to introduce a child to Christ? Our opportunity to help people in economic distress will be limited only by our own imagination or by the expansiveness of our hearts.

Whatever ministry opportunities we choose to consider, we should lay them before the Lord and seek His will and His blessing. And we should start now, because every day there is less time to labor for the Lord than there was the day before. Time is so short. The stern warning we received more than 100 years ago is more urgent now than it ever was:

"The days in which we live are solemn and important. The Spirit of God is gradually but surely being withdrawn from the earth. Plagues and judgments are already falling on the despisers of the grace of God. The calamities by land and sea, the unsettled state of society, the alarms of war, are portentous. They forecast approaching events of the greatest magnitude.

"The agencies of evil are combining their forces, and consolidating. They are strengthening for the last great crisis. Great changes are soon to take place in our world, *and the final movements will be rapid ones.*"[16]

If the worthies who pioneered the Adventist Church and who were so versed in prophecy were alive today, I truly feel that they would go to any length and any embarrassment—including standing on street corners with signs—to get our attention and wake us from our spiritual sleep. There is so little time, and the events coming upon us are so enormous and overpowering, that we must seek the Lord with all our heart and do His will with all our strength, not letting any earthly consideration stand between us and our Savior.

---

[1] E. G. White, *God's Amazing Grace,* p. 100.

[2] Adventist Book Centers can help with your search.

[3] See E. G. White, *Country Living,* for an excellent resource on this question.

[4] E. G. White, *Manuscript Releases,* vol. 1, p. 248.

[5] Ellen G. White, *Testimonies for the Church* (Mountain View, Calif.: Pacific Press Pub. Assn., 1948), vol. 5, pp. 464, 465.

[6] E. G. White, *Country Living,* p. 17.

[7] Such as prayer and meditation, Bible study, study of inspired writings, gardening, service to others.

[8] Ellen G. White, *Humble Hero* (Nampa, Idaho: Pacific Press Pub. Assn., 2009), p. 137.

[9] E. G. White, *Prophets and Kings,* p. 626.

[10] E. G. White, *The Great Controversy,* pp. 593, 594.

[11] E. G. White, *Testimonies for the Church,* vol. 4, p. 459.

[12] Ellen G. White, *The Ministry of Healing* (Mountain View, Calif.: Pacific Press Pub. Assn., 1905), p. 512.

[13] E. G. White, *Testimonies for the Church,* vol. 4, p. 538.

[14] Ellen G. White, *Steps to Christ* (Mountain View, Calif.: Pacific Press Pub. Assn., 1956), p. 101.

[15] I have witnessed the effectiveness of this ministry at the Barre, Vermont, church.

[16] E. G. White, *Testimonies for the Church,* vol. 9, p. 11. (Italics supplied.)

Chapter 11:

It's Not *What* Is Coming;
It's *Who* Is Coming That Matters

"The nations are in unrest. Times of perplexity are upon us.
Men's hearts are failing them for fear of the things that are coming
upon the earth. But those who believe in God will hear His
voice amid the storm, saying, 'It is I; be not afraid.'"
—*Ellen G. White,* Signs of the Times, *October 9, 1901.*

I can hardly wait for Christ to come. The thought of meeting my friend Jesus and of seeing the unending delights of heaven thrills me. The thought of being in a place where there is no sin, no sickness, no suffering, no decay, no deformity, no death—this absolutely makes my heart sing. For me, heaven is a very real place that I long for. Aside from all the obvious reasons to long for heaven, I guess I should mention here that one of my sons, Nathan (who is 20 years old at this writing), has Down syndrome. Nathan and I are buddies, and not only do I enjoy his company, but I also respect him for the young man he is and for the hurdles he tries to (and sometimes does) overcome. Even so, my heart aches to see him transformed in the twinkling of an eye, to see him without spot or stain of sin, to have an actual extended conversation with him. I live daily in anticipation of meeting my Savior and of seeing my son transformed, not into just a "normal" person, but what God intended him to be in a sinless world. I also admit to a great deal of curiosity as to what I will be like when transformed to a state that is not marred by sin. I am sure the transformation will be every bit as large for me as it will be for my son Nathan.

The difference between humans as we were created and humans as we now exist is very large, and we can assume that the same is true for the earth. When God created the earth, all of heaven was amazed and sang with joy. It was that good. *The Desire of Ages* quotes Job 38:7 in describing the response of the heavenly hosts to the completed creation: "The morning stars sang together, and all the sons of God shouted for joy."[1] But sin is in the advanced stages of destroying that once-perfect world. God gives a clue as to how He feels about that in Revelation 11:18: "And the nations were angry, and thy wrath is come, and the time of the dead, that they should be judged, and that thou shouldest give reward unto thy servants the prophets, and to the saints, and them that fear thy name, small and great; and shouldest destroy them which destroy the earth."

From the laying of the foundations of the earth, God had a contingency plan for redeeming humanity and for redeeming His world. *Patriarchs and Prophets* expressly underscores the point that the earth—the physical earth—is part of the redemption plan: "Not only man but the earth had by sin come under the power of the wicked one, and was to be restored by the plan of redemption."[2] And now, as history winds down, we are much nearer the day on which this earth will be destroyed, and a new, perfect earth will be created in its place. This is extremely exciting news for anyone who has ever found joy in the beauty of this earth, or admired life in its many forms—from land animals to sea creatures to trees and grass and plants. The earth will be re-created without any trace of sin and will be such an amazing, vibrant, joyful place that we will look back on this earth as being but a dim shadow of what it was meant to be. Right now we have only the faintest idea of the beauty, joy, and celebration that awaits us in heaven and on the earth made new.

A new earth is *what* is coming, but that is of little import compared to *who* is coming. In that context, here is the critical question for you: Is Christ's coming a present reality for you? If so, is your personal relationship with Christ the paramount driving force in your life? Or do you settle for a lackluster and impersonal relationship with Christ while you walk with the world?

Do you like a really good movie or a really good book—the kind that has twists and turns in its plot so that you don't really know how it is going to end? I admit to a fondness for such a plot, especially the kind in which the bad guys think they are in control, and then there is a big twist and a turn, and *wham!* The bad guys end up being shut out and slowly realize they have been played. What they thought was real turned out to be an illusion.

The greatest story ever told, unlike movies and books, doesn't have that many twists and turns. If you read your Bible, you will know there are no secrets, no sudden twists in the plot—how it all turns out is there for everyone to see well ahead of time. And if you read the Spirit of Prophecy, you will know all the significant landmarks that are to take place in these last days, so that, between the two, you know what is happening and what will happen. And yet, when the big, final scene at the end of the story of this earth is played out, and Christ is seen in the clouds coming for His faithful, how many slack-jawed, dumbfounded people will suddenly realize that they have been played?

How about you? Are the tangible temptations of this world your reality? Money, power, respect, cool stuff? Is that what is real to you? If so, when the sky rolls away as if it were a scroll, when clouds of angels fill your field of view and it becomes starkly obvious that nothing in this earth is of any real, lasting value, will you be saying, "I got played"? Or will you be among those who are ecstatic and joyfully shouting and who are looking forward to an endless life in a place that was created specifically to delight us? What is there on this earth that is fair compensation for missing out on heaven?

The world that God originally created was perfect. Who wrecked it? If your answer is that Satan wrecked it, then, in my opinion, you are wrong. Satan could not have done anything to this world without human beings willfully sinning. So in truth, it was humanity who disobeyed God and started the destruction of His earth. And it is humanity who has hastened this decay and deformity and death throughout history, in the process becoming more and more like Satan. By any human standard we are guilty, and the fact that everything God created on the earth has suffered and has either died or is dying is our fault. By any human standard we should be punished by death. But, praise God, it is not human standards by which we are being judged.

Here, brothers and sisters, is the really good news: Instead of giving us the swift and severe punishment that we deserved, God sacrificed His own Son to pay the penalty in our place. Think on that for a moment. We wrecked His planet, rejected Him, scorned His Son and His prophets, and in return He offers us the chance to be forgiven and experience everlasting love and delight in a perfect new world. This, in a nutshell, tells us everything we need to know of the character of God and also of the character of Christ, who willingly let Himself be sacrificed. And soon—very, very soon—this same Christ is coming back to this earth to claim those who claim Him. In my opinion He is coming within our lifetimes, and His coming is the reality that we should be continuously focused on. Nothing else matters.

Let me ask you something: Are you willing to be embarrassed for Christ?

Planet in Distress

So many Christians claim they are willing to die for Christ if needed, or are willing to work wherever they are called, or are willing to sacrifice all of their possessions, etc. That is all well and good. But in the final analysis, how many Christians are willing to suffer embarrassment in the eyes of the world by doing something like going door to door and handing out literature, or asking a neighbor if they can pray together, or by undertaking any other act that requires us to openly take a stand with Christ and be seen as "weird" in the eyes of the world? Unfortunately, while we claim to be willing to honor and glorify Christ through almost any sacrifice, so many of us are in fact unwilling to take a public stand for Him and work for Him because of what the world may think.

It is time we get over our embarrassment of our Savior and settle into our role of being a "peculiar people" who proclaim Christ openly and joyfully. Christ says in Matthew 10:32, 33: "Therefore whoever confesses Me before men, him I will also confess before My Father who is in heaven. But whoever denies Me before men, him I will also deny before My Father who is in heaven" (NKJV). It is time we distance ourselves sufficiently from the world that such concerns as pride do not interfere with our relationship with Christ. It is time that we see the real priorities that are important to God in these last days and that we reorder our lives as necessary to undertake these priorities while building a close, personal relationship with Christ. And we need to do all of this starting now—today.

We are told that the last events will be rapid ones. Based on what this book has explored, we know that irresistible forces are even now at work that will bring about the condition of the world in the last days as described by Christ Himself. We can actually see and measure the progression of the world toward its end. If there was ever a time to be proclaiming Christ— loudly and without embarrassment or hesitation—it is now.

Jesus really is coming, and He really is coming very, very soon.

---

[1] E. G. White, *The Desire of Ages,* p. 769.
[2] E. G. White, *Patriarchs and Prophets,* p. 67.

Our Hope for the Future

How to Survive Armageddon
John C. Brunt

Professor John C. Brunt opens the Bible to give you an intriguing glimpse into the future. You'll see deception, war, plagues, pestilences, and persecution. But there is also something that will give you hope, confidence, and even joy! 978-0-8280-2582-9.

Uplifting Stories
of God's Enduring Promises

It's Going to Be All Right
Arthur A. Milward

Each of these poignant stories includes a thread of hope—our heavenly Father has assured us that in the end everything will be all right. In the meantime, though, His promises soothe our aching souls. 978-0-8280-2563-8. **US$11.99** eBOOK AVAILABLE

Judgment
and Assurance

The promise of salvation is assured for those who trust in the Savior.

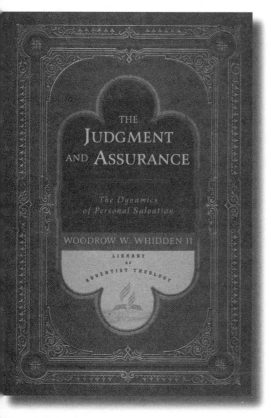

The Judgement and Assurance
Woodrow W. Whidden II

Inspiring and instructive, this book addresses common questions and misunderstandings about the assurance of salvation, grace, law, judgment, and final end-time events.
Hardcover 978-0-8280-2565-2.